BIOCHEMISTRY RESEARCH TRENDS

A CLOSER LOOK AT THE COMET ASSAY

BIOCHEMISTRY RESEARCH TRENDS

Additional books and e-books in this series can be found on Nova's website under the Series tab.

BIOCHEMISTRY RESEARCH TRENDS

A CLOSER LOOK AT THE COMET ASSAY

KEITH H. HARMON
EDITOR

Copyright © 2019 by Nova Science Publishers, Inc.

All rights reserved. No part of this book may be reproduced, stored in a retrieval system or transmitted in any form or by any means: electronic, electrostatic, magnetic, tape, mechanical photocopying, recording or otherwise without the written permission of the Publisher.

We have partnered with Copyright Clearance Center to make it easy for you to obtain permissions to reuse content from this publication. Simply navigate to this publication's page on Nova's website and locate the "Get Permission" button below the title description. This button is linked directly to the title's permission page on copyright.com. Alternatively, you can visit copyright.com and search by title, ISBN, or ISSN.

For further questions about using the service on copyright.com, please contact:
Copyright Clearance Center
Phone: +1-(978) 750-8400   Fax: +1-(978) 750-4470   E-mail: info@copyright.com.

NOTICE TO THE READER

The Publisher has taken reasonable care in the preparation of this book, but makes no expressed or implied warranty of any kind and assumes no responsibility for any errors or omissions. No liability is assumed for incidental or consequential damages in connection with or arising out of information contained in this book. The Publisher shall not be liable for any special, consequential, or exemplary damages resulting, in whole or in part, from the readers' use of, or reliance upon, this material. Any parts of this book based on government reports are so indicated and copyright is claimed for those parts to the extent applicable to compilations of such works.

Independent verification should be sought for any data, advice or recommendations contained in this book. In addition, no responsibility is assumed by the Publisher for any injury and/or damage to persons or property arising from any methods, products, instructions, ideas or otherwise contained in this publication.

This publication is designed to provide accurate and authoritative information with regard to the subject matter covered herein. It is sold with the clear understanding that the Publisher is not engaged in rendering legal or any other professional services. If legal or any other expert assistance is required, the services of a competent person should be sought. FROM A DECLARATION OF PARTICIPANTS JOINTLY ADOPTED BY A COMMITTEE OF THE AMERICAN BAR ASSOCIATION AND A COMMITTEE OF PUBLISHERS.

Additional color graphics may be available in the e-book version of this book.

Library of Congress Cataloging-in-Publication Data

ISBN: 978-1-53611-028-9
Library of Congress Control Number:2019950410

Published by Nova Science Publishers, Inc. † New York

# CONTENTS

Preface		vii
Chapter 1	Clinical Applications of Comet Assay *Merve Bacanli*	1
Chapter 2	Comet Assay in Occupational Toxicology Studies *Hatice Gül Anlar*	17
Chapter 3	The Comet Assay as a Tool to Detect the Genotoxic Potential of Nanomaterials *Constanza Cortés and Ricard Marcos*	35
Chapter 4	Kinetic Approach in Comet Assay: An Opportunity to Investigate DNA Loops *Katerina Afanasieva and Andrei Sivolob*	65
Chapter 5	Evaluation of Global DNA Methylation Status of Single Cells by the Comet Assay: A Promising Approach in Cancer Diagnosis and Follow-Up *Yildiz Dincer*	85

Chapter 6	Comet Assay: A Suitable Method for *in Vitro* Genotoxicity Assessment Using Animal Lymphocytes *Simona Koleničová, Viera Schwarzbacherová, Beáta Holečková and Martina Galdíková*	105
Chapter 7	Determination of Aluminum-Induced Oxidative and Genotoxic Effects in Sunflower Leaves *Aslıhan Çetinbaş-Genç, Elif Kılıç-Çakmak, Fatma Yanık, Filiz Vardar, Ahu Altınkut-Uncuoğlu and Yıldız Aydın*	143
Chapter 8	Determination of Genotoxic Effects of Organochlorine Pesticides in Wheat (*Triticum aestivum* L.) by Comet Assay *Melek Adiloğlu Öztürk and Yıldız Aydın*	171
Chapter 9	Investigation of Genotoxic Effects of Organophosphorus Pesticides in Barley (*Hordeum vulgare* L.) *Tuba Akan and Yıldız Aydın*	187
Chapter 10	Genotoxicity of Hydroquinone and Fungal Detoxification: Correlation with Hydroquinone Concentration and Cell Viability *Ana Lúcia Leitão*	203
Index		**221**
Related Nova Publications		**229**

PREFACE

A Closer Look at the Comet Assay opens with a discussion on the clinical applications of comet assay. Comet assay is rapid, simple method which able to assess DNA damage in different samples like blood, cells and tissues.

Following this, the authors examine comet assay usage in occupational toxicology studies. Isolated lymphocytes were the most used cell line in these studies, but exfoliated cells such as nasal and buccal cell, liver, kidney and sperm cells may be used.

Comet assay may also be used to detect nanoparticles-associated DNA damage. As such, this compilation assesses potential limitations due to the interaction of the nanoparticles with the method.

Next, to shed light on the mechanisms of the DNA track formation, the authors apply an original approach based on the kinetic measurements in the comet assay, arguing that in neutral conditions at low levels of DNA damages, the comet tail is formed by extended DNA loops.

New applications of the comet assay are described for the detection of aberrant DNA methylation, which is a promising marker in cancer diagnosis and follow-up.

The authors go on to describe and analyse the results of in vitro treatment of lymphocytes with insecticide using comet assay under alkaline and neutral conditions, testing the commercial product Calypso®

480SC and its active agent thiacloprid at concentrations of 30; 60; 120; 240 and 480 µg.ml-1.

In one study, Helianthus annuus (sunflower) seedlings were irrigated with Hoagland solution containing different concentrations of $AlCl_3$. Morphological parameters such as germination rate and stoma number are evaluated.

Additionally, the genotoxic effects of endosulfan pesticide at different times and in different concentrations in wheat leaf samples are analyzed in two-week old wheat seedlings in an effort to demonstrate that endosulfan is a genotoxic agent causing DNA breaks in wheat.

In the closing chapter, the correlation between the comet assay parameters, cell viability, and hydroquinone concentration is explored. The relationship between comet assay and remaining hydroquinone after fungal treatment is also investigated in order to evaluate its biodegradation efficiency.

Chapter 1 - The cells in the human body are exposed to various endogenic and exogenic toxic substances, which have the potency to cause DNA damage. DNA damage may result in chronic diseases such as diabetes, cardiovascular diseases, Alzheimer disease, cancer and aging. Because of this, the detection of DNA damage is very important. Comet assay is rapid, simple method which able to assess DNA damage in different samples like blood, cells and tissues. This assay should be suitable for use in the clinical application which requires only a few cells, and results can be obtained quickly. In this chapter, the clinical applications of Comet assay will be discussed.

Chapter 2 - The comet assay, also known as single cell gel electrophoresis (SCGE), is a gel electrophoresis method used to visualize and measure DNA strand breaks in individual cells, using microscopy. It has been found to be a very sensitive, rapid, reliable and fairly inexpensive way of measuring DNA damage. It has a further advantage that the observations are made at the single cell level. It has been increasingly used in occupational biomonitoring studies with different occupational settings. Isolated lymphocytes were the most used cell line in these studies but also exfoliated cells like a nasal and buccal cell, liver, kidney and sperm cells can be used. For example, according to the authors' previous studies with ceramic workers and welders, the results showed that these workers had significantly more DNA damage in their isolated lymphocytes and whole blood compared to controls and also, DNA damage in lymphocytes and whole blood cells were correlated. It was important because using whole blood is simpler since the isolation of lymphocytes requires more time and chemicals. This chapter aims to provide knowledge about Comet assay and its usage in occupational toxicology studies.

Chapter 3 - Since their discovery in the 1980s, the use of nanoparticles (NPs) has grown exponentially due to their distinctive physicochemical properties, which are exploited in fields as broad as electronics, medicine, food production, and packaging. However, this brings into question the potential toxicological issues derived from the increased exposure to NPs, especially DNA damage. It should be pointed out that, according to the relevance of the target, genotoxicity can determine undesired consequences such as immune responses or carcinogenicity. Out of the numerous assays utilized to detect NPs genotoxicity, the comet assay, an electrophoretic technique allowing the detection of DNA strand breaks in single cells, is the most commonly used. This chapter focuses on its use to detect NPs-associated DNA damage, as well as its potential limitations due to the interaction of the NPs with the method. To achieve these goals, the authors present an overview of the available literature where this assay has been used to test NPs genotoxicity.

Chapter 4 - The comet assay is thought to be a sensitive and simple technique to assess DNA single- and double-strand breaks at the level of individual cells. The approach is based on an analysis of parameters of the electrophoretic track (the comet tail) formed during electrophoresis of nucleoids obtained after cell lysis in agarose on a surface of microscopic slide. The physical principles of DNA migration in this electrophoretic system remained to be rather elusive for a long time. To shed light on the mechanisms of the DNA track formation the authors applied an original approach based on the kinetic measurements in the comet assay. This chapter is focused on the authors' recent results, which allows them to argue that in neutral conditions at low levels of DNA damages (and also in the case of undamaged cells) the comet tail is formed by extended DNA loops and, more important, these loops are about the same as chromatin loops in the cell nuclei. The authors' approach gives an opportunity to investigate several parameters of the loops including the loop supercoiling and large-scale features of the loop domain organization (and reorganization) in nucleoids obtained from cells of different types.

Chapter 5 - As an epigenetic modification, DNA methylation plays a pivotal role in gene regulation, and is crucial for maintaining genome stability. DNA methylation occurs generally in cytosine within CpG dinucleotides forming 5-methylcytosine (5-mC) on gene promoters and is associated with repression of transcription. This epigenetic event is reversible, removal of methyl group causes transcription of the gene. DNA methylation patterns can be established on a global or gene-specific level in accordance with regulatory needs of cells. If the DNA methylation process is not correctly regulated, it can lead to disruption in gene expression, impairment in biological pathways and finally development of a wide range of diseases. Aberrant DNA methylation is the most common feature of tumor cells. Genomic instability, oncogene activation and tumor suppressor gene inactivation through altered DNA methylation patterns are well defined in tumor cells. DNA methyltransferase inhibitors, which are designed to demethylate DNA, have already been introduced into clinical use for cancer therapy. In order to detect DNA methylation signature in cancer diagnosis and prognosis, various methods have been established. DNA sequencing after bisulfite conversion is the most frequently used assay to measure both global and gene-specific DNA methylation. However, all methods are expensive, require long time and there is no standardization and unification. Comet assay is a method for assessment of DNA damage and repair quantitatively at the level of single cell. The assay has been used in testing genotoxicity of chemicals, monitoring environmental contamination with genotoxic compounds, human biomonitoring and molecular epidemiology. In addition to DNA strand breaks, some base lesions can be measured by the comet assay using lesion-specific endonucleases. In this context, comet assay is frequently used to determine oxidative DNA damage using 8-oxoguanine glycosylase 1. Recently, a new modification of the comet assay was established to determine the global DNA methylation level of individual cells using 5-methylcytosine (5-mC)-specific restriction endonucleases that are HpaII, MspI and McrBC. The EpiComet-Chip, a modified high-throughput comet assay for the determination of DNA methylation status was introduced. In this chapter, this new application of the comet assay is decribed for detection of aberrant DNA methylation which is a promising marker in cancer diagnosis and follow-up.

Chapter 6 - Neonicotinoids make up a widely-used group of insecticides which are under increased regulatory control because of their harmful impact on non-target organisms. Sensitive, simple and rapid test systems are required for pesticide risk assessment and biomonitoring studies. The *in vitro* Comet assay is now recommended as a suitable test in the technical guidance documents for the Registration, Evaluation and Authorisation of Chemicals (REACH), and is widely used in genotoxicity testing of pesticides, nanoparticles and pharmaceuticals. Comet assay under alkaline (DNA single- and double-strand break detection) or neutral (DNA double-strand break detection) conditions has been widely performed on cultured cells or cell lines for DNA damage assessment. DNA damage is frequently used as a biomarker connected with exposure, the process of ageing, diseases and cancer development. Here the authors describe and analyse the results of *in vitro* treatment of lymphocytes with insecticide using Comet assay under alkaline and neutral conditions. The authors tested the commercial product Calypso® 480SC and its active agent thiacloprid at concentrations of 30; 60; 120; 240 and 480 μg.ml^{-1}. The experiments were performed on isolated bovine lymphocytes, which were treated with insecticide for 2 h in two different ways: immediately after isolation and for the last 2h of 48h cultivation in cell culture media. Cattle and ruminants represent a suitable experimental model for genotoxicity assessment because of their higher exposure to pollutants than other animals through their diet.

First the impact of commercial product Calypso® 480SC was evaluated. Results of *alkaline Comet assay* showed significantly elevated DNA damage in the concentration range from 60 to 480 μg.ml^{-1} ($p < 0.01$

and $p < 0.001$) after 2h treatment. The experiments done with pre-cultivation of lymphocytes before 2h treatment produced an increase in DNA strand breaks ($p < 0.05$) only at the highest tested concentrations (240 and 480 µg.ml^{-1}). *Neutral Comet assay* showed statistical significance only at the highest concentration tested (480 µg.ml^{-1}, $p < 0.05$) in both procedures, with and without pre-cultivation of lymphocytes. Next the authors analysed the results after exposure to pure active agent thiacloprid. Under *alkaline* conditions, increased levels of DNA strand breaks were detected at the highest concentration tested (480 µg.ml^{-1}; $p < 0.05$ and $p < 0.05$) in the basic and pre-cultivation procedure. *Neutral* Comet assay did not record any statistically-significant extent of DNA damage at any of the concentrations tested. The authors' results suggest that commercial insecticide is able to alter genetic material, and Comet assay was found to be appropriate for early detection of DNA damage.

Chapter 7 - According to rapid industrialization and urbanization serious ecological problems come into prominence worldwide. Agronomic crops and wild flora are faced with environmental risk factors primarily inducing reduced crop quality and yield due to their sessile nature. Most of the economic losses are related to soil acidification and heavy metal accumulation causing adverse effects on plant growth and development. Aluminum (Al) is one of the most abundant elements in the earth crust comprising 8.1%. It exists in the form of insoluble aluminosilicates or oxides commonly. If the soil pH reduces (pH <5.0), the complex Al is dissolved and absorbed by plant root system which is the first target organ in plants. Since the root apex accumulates more Al, Al toxicity represents its adverse effects on root growth initially resulting in low water and nutrient uptake. Because of being the first target organ, most of the Al toxicity researches subjected root growth and development, whereas limited studies are available on foliar symptoms of Al toxicity. In the present study *Helianthus annuus* (sunflower) seedlings were irrigated with Hoagland solution containing with or without different concentrations of AlCl$_3$ (50, 100, 150 and 200 µM pH 4.5). After eight weeks, fresh leaves were analyzed for determining the oxidative and genotoxic responses. Morphological parameters such as germination rate and stoma number were evaluated. Antioxidant enzyme analyses (superoxide dismutase, catalase and peroxidase activity) were performed in sunflower leaves for determination of oxidative stress. Total chlorophyll, carotenoid and anthocyanine content were also measured. The genotoxic effects of Al were performed by comet assay in sunflower leaves. It has been known that comet assay is a practical and sensitive implementation to assess the genotoxic impact of various type of stress factors such as pesticides, UV and heavy metals. The assay is based on the quantification of denatured DNA fragments which migrates out of the nucleus during electrophoresis. According to the authors' results, Al caused adverse effects on sunflower leaves in all concentrations. Although the toxicity level was dose dependent up to 150 µM, it reduced at 200 µM in compare to 150 µM. The comet assay results also revealed that Al induced DNA damage confirmed by increase in % DNA tail, olive tail moment and tail intensity. The DNA damage was evident in sunflower leaves which is not the first target as roots. In conclusion, sunflower leaves showed a resistance up to 150 µM Al by enhancing its stress tolerance mechanism, but 200 µM Al was over dose that blocked its balance.

Chapter 8 - Due to the detection of pesticides in underground and surface waters in many agricultural regions of the world, the environmental dimension of pesticide use is an important issue discussed today. Continuous use of pesticides in agriculture causes genetic resistance to pests. Pesticides threaten the ecosystem by altering the structure and species distribution of the ecosystem and disrupting the normal balance between food chains. Therefore pesticides need to be monitored under strict supervision.

In this study, the authors' aim was to determine the genotoxic effects of endosulfan (ES) pesticide at different time and concentrations in wheat (*Triticum aestivum* L.). Leaf samples were taken from two-weeks old wheat seedlings of the negative control (Hoagland solution), positive control (Hoagland solution containing 0.1% H2O2) and different concentrations (1 g/L, 2 g/L and 4 g/L) of ES treated groups at different time intervals (6h, 12h and 18h) and examined by comet technique (single cell gel electrophoresis).

DNA damage levels increased significantly after the ES application in wheat seedlings time and dose dependently evaluated by comet analysis. In wheat seedlings, the highest DNA damage (% Tail DNA:50, Olive tail moment: 0.34) were determined at 12h of 2 g/L ES concentration ($p < 0.001$).

The obtained results of this study demonstrated that the ES is a genotoxic agent causing DNA breaks in wheat. Also, the results related with decrease in the level of DNA damage obtained from this study will contribute to the determination of the spraying and harvesting time as well as the determination of the appropriate concentrations using this pesticide.

Chapter 9 - Chlorpyrifos [O O-diethyl-O-(3 5 6-trichloro-2-pyridyl)-phosphorothioate (CPF)] is one of the most widely used organophosphate insecticides. Pesticides are increasingly used chemicals to enhance the yield and quality of agricultural products. With their increasing use, they pose a risk for both human health and ecosystem. While pesticides used against harmful effects of agricultural production affect the target organisms, they can affect the non-target organisms through wrong and especially excessive applications. Plant communities are often subjected to target as well as non-target exposure to pesticides. The Comet assay is used as a useful tool in assessing the potential of plants as source of information on the genotoxic impact of dangerous pollutants and sensitive sensors in ecosystems. This technique is based on the quantification of denatured DNA fragments migrating out of the cell nucleus during electrophoresis. This research was performed to determine the genotoxicity potential of Chlorpyrifos (CPF) at different concentrations and exposure times in barley by comet assay. The barley seedlings were treated with negative control (Hoagland's solution), positive control (Hoagland's solution with %0,1 H_2O_2) and different concentrations (5ml/L, 10ml/L, 20ml/L and 40 ml/L) of CPF for 6, 12 and 18h. In barley seedlings, the highest single stranded DNA breaks were determined for (%DNA tail: 55, Olive tail moment: 0.36) 6h of 5ml/L CPF dosage. During the exposure time, although the linear decrease in DNA damage has been observed in leaves of barley, the DNA damage was increased in 40ml/L CPF for 12 h (% DNA tail: 47, Olive tail moment: 0.29). The authors' research findings clearly indicated that CPF has genotoxic effects in barley. The findings may contribute to the determination of the appropriate doses for the use of ES, as well as the determination of spraying and harvesting time.

Chapter 10 - Hydroquinone, the major benzene metabolite, is an occupational and environmental pollutant. Hydroquinone can be produced as a result of human activities and industrial processes, as well as during phenolic and benzene biotransformation. Additionally, it can be auto-oxidized to form a product, 1,4-benzoquinone, that has higher toxicity than the parent compound. Among fungal strains, the halotolerant *Penicillium chrysogenum* var. *halophenolicum* is a versatile microorganism for hydroquinone biotransformation. The Presto blue® dye assay is a biologically safe and sensitive test for cytotoxic assessment. In the present study, the correlation between the Comet assay parameters, cell viability, and hydroquinone concentration was performed. It was also investigated the relationship between Comet and remaining hydroquinone after fungal treatment in order to evaluate its biodegradation efficiency. There was a high correlation between % DNA in tail, olive tail moment, tail length and percentage of survival of human colon cancer cells, HCT116, and hydroquinone concentrations. While a positive correlation between Comet parameters and hydroquinone concentrations was achieved; there was a negative correlation between Comet parameters and percentage of survival. *Penicillium chrysogenum* var. *halophenolicum* reduced DNA damage due to hydroquinone degradation. The same trend was observed for experiments under hyperosmolality, reinforcing *Penicillium chrysogenum* var. *halophenolicum* potential for hydroquinone remediation.

In: A Closer Look at the Comet Assay
Editor: Keith H. Harmon

ISBN: 978-1-53611-028-9
© 2019 Nova Science Publishers, Inc.

Chapter 1

CLINICAL APPLICATIONS OF COMET ASSAY

Merve Bacanli[*]

Gülhane Pharmacy Faculty Department of Pharmaceutical Toxicology,
University of Health Sciences, Ankara, Turkey

ABSTRACT

The cells in the human body are exposed to various endogenic and exogenic toxic substances, which have the potency to cause DNA damage. DNA damage may result in chronic diseases such as diabetes, cardiovascular diseases, Alzheimer disease, cancer and aging. Because of this, the detection of DNA damage is very important. Comet assay is rapid, simple method which able to assess DNA damage in different samples like blood, cells and tissues. This assay should be suitable for use in the clinical application which requires only a few cells, and results can be obtained quickly. In this chapter, the clinical applications of Comet assay will be discussed.

Keywords: genotoxicity, comet, biomonitoring, DNA damage, occupational genotoxicity

---

[*] Corresponding Author's Email: mervebacanli@gmail.com.

1. INTRODUCTION

There are various toxic factors in the environment which have oxidative potential to cause DNA damage in human body cells (Kryston et al. 2011). Every day, human populations are exposed to mutagenic and carcinogenic compounds, both occupationally and/or environmentally (Azqueta et al. 2014).

Genetic biomonitoring of populations exposed to mutagens and/or carcinogens is an important warning system for genetic diseases and cancer. It also allows identification of risk factors. Human genetic biomonitoring studies include various cytogenetic markers including single cell gel electrophoresis (Comet), chromosomal aberration, micronucleus and sister chromatid exchange (Kassie, Parzefall, and Knasmüller 2000).

Comet assay is a useful tool for determining low levels of exposure to genotoxins (Collins, Dušinská, et al. 1997). This assay is also important for determining protective effects of antioxidants and their DNA repair capacity in various diseases (Gunasekarana, Raj, and Chand 2015). In comparison with other genotoxicity methods, this assay is relatively robust and economical in its use of material. In Comet assay, data is collected at the level of the individual cells, only a small number of cells required, and data can be obtained within a few hours of sampling (Rojas, Lopez, and Valverde 1999).

This chapter aims to provide knowledge about the general information about Comet assay and its applications.

2. COMET ASSAY

The Comet assay was developed as a method to detect DNA damage and also DNA repair. Single and double strand breaks, apurinic/apyrimidinc (AP) sites and repair intermediates can be detected by Comet assay (Gunasekarana, Raj, and Chand 2015). In its protocol, cells are embedded in agarose, placed on a microscope slide and lysed to remove membranes and soluble components with hypertonic, non-ionic detergent solutions. After that, nucleoids are exposed to neutral or alkaline treatment and electrophoresis. The presence of breaks in the DNA relaxes the supercoiled loops and enables the DNA to migrate towards the anode. Finally, DNA is stained with a fluorescent dye and visualized by fluorescence microscope. The digestion of the nucleoids with lesion specific enzymes such as formamidopyrimidine DNA glycosylase (Fpg) and 8-oxo-7,8-dihydroguanine (8-oxoGua), endonuclease III, T4 endonuclease V allows the other lesions such as oxidized bases (Azqueta et al. 2019).

A variety of normal and transformal cells of human, animal and plant have been used for different *in vitro* studies including Comet assay. Most of the studies have used leukocytes and lymphocytes but other tissues including epithelial, reproductive, pancreatic cells and different human cell lines such as cancer and fibroblast cells have been also used (Rojas, Lopez, and Valverde 1999). Collecting blood or tissues is not always feasible in all human subjects. Therefore, the other sources of cells that can be collected via non-invasively route have been used for Comet assay including epithelial cells as well as sperm (Table 1) (Cortés-Gutiérrez et al. 2014, Rojas et al. 2014, Langie, Azqueta, and Collins 2015).

Comet assay in combination with fluorescent labeled targeted against particular DNA sequences constitute Comet-FISH assay, which has been used to assess the DNA damage and repair of single genes or short DNA sequences. It is known that Comet-FISH assay can be used with polymerase chain reaction (PCR). Apo-necro Comet assay, another type of Comet assay, can be used to differentiate viable, apoptotic and necrotic cells (Gunasekarana, Raj, and Chand 2015).

Table 1. Human tissues from which cells can be used in Comet assay

Nasal tissue
Buccal tissue
Blood
Bladder tissue
Colon tissue
Prostate tissue
Testis tissue

3. APPLICATIONS OF COMET ASSAY

3.1. *In Vitro* Applications

There are various studies about *in vitro* genotoxicity in different cell lines. The effects of different polyphenols on the base excision repair (BER) activity of rat pheochromocytoma cell line (PC12) were determined. A significant increase was found in the incision activity of cells treated with rosmarinic acid (Silva, Gomes, and Coutinho 2008). In a different study, incubation of human cervical cancer cells (Caco-2) with water extracts of *Salvia* species, luteonil-7-glucoside and rosmarinic acid increased BER activity of cells (Ramos et al. 2010). Azqueta et al. (2013) also showed that vitamin C caused DNA breaks in nucleoids in Caco-2 cells (Azqueta et al. 2013). In the human lymphocyte and Chinese hamster fibroblast (V79) cells, it is found that galangin, puerarin, ursolic acid, limonene and naringin revealed a reduction in the DNA damage caused by hydrogen peroxide (H_2O_2) (Bacanlı, Başaran, and Başaran 2015, 2017).

3.2. Animal Studies

Various animal models such as fish, rats and mice have been used to assess the DNA damage because of different genotoxicants and disease models by Comet assay (Singh 2018).

In a study, male and female mice repeatedly irradiated with X-rays and injected with nonylphenols for 2 weeks, 5 days in a week. Nonylphenol induced DNA damage was seen in different tissues of mice in Comet assay (Dobrzyńska 2014). In another study, Zhang et al. (2014) examined the genotoxic effects of flumorph in the brain, liver, spleen, kidneys and sperms of mice and found significant increases in the DNA damage in a dose and time dependent manner. In rodents, smoke induced DNA damage was seen in isolated cells following single or 5-day smoke exposure in Comet assay (Dalrymple et al. 2015).

In streptozotocin induced diabetic rats, the protective effects of limonene, pycnogenol®, ursolic acid and cinnamic acid against the DNA damage caused by diabetes were shown in the blood, liver and kidney samples (Bacanlı et al. 2017, Bacanlı et al. 2018, Bacanli et al. 2018, Anlar et al. 2018, Aydın et al. 2019). Besides, it is reported that ferulic acid, pycnogenol®, rosmarinic acid and resveratrol might have a role in the attenuation of sepsis-induced oxidative damage not only by decreasing the DNA damage but also by increasing the antioxidant status and DNA repair capacity of the animals (Aydın et al. 2013, Taner et al. 2014, Bacanlı et al. 2014, Bacanlı et al. 2016, Aydın et al. 2016). Tokaç et al. (2017) also demonstrated the ameliorative effects of pycnogenol® on liver ischemia reperfusion injury induced DNA damage in rats.

3.3. Occupational Studies

In the textile, ceramic, welding, dyeing, iron, steel and leather industries, the workers expose to various fumes, metal particulates, gases, dyes and solvents which are genotoxic (Singh and Chadha 2014). Most of the metal particulates, dyes and solvents can enter the cells and cause single and/or double DNA breaks (Shah, Lakkad, and Rao 2016). Those effects are generally subclinical, unless cumulative changes lead to chronic health effects including cancer, cardiovascular and neurodegenerative diseases after long-term exposure (Singh 2018). Comet assay was shown to be able to identify low levels of exposure with great sensitivity (Kassie, Parzefall, and Knasmüller 2000).

Bhalli et al. (2006) assessed DNA damage in the leukocytes of Pakistani pesticide factory workers who exposed to organophosphates, carbamates and pyrethroids. The workers were found to have had significantly higher DNA damage in Comet assay. Megyesi et al. (2014) studied the genotoxic effects of benzene, polycyclic aromatic hydrocarbons (PAH) and styrene in exposed groups with Comet assay. In a study with bakery workers, the DNA damage of workers were found significantly higher than controls (Kianmehr, Hajavi, and Gazeri 2017).

The DNA damage of Turkish asphalt workers were detected in the lymphocytes by Comet assay. Results demonstrated that workers had more DNA damage when compared to their controls (Bacaksiz et al. 2014). In a different study, the genotoxic potential of perchloroethylene (PCE) was examined in dry-cleaning workers. The DNA damage evaluated by Comet assay was significantly lower in controls than the workers (Everatt et al. 2013).

Anlar et al. (2017) were concluded that occupational chemical mixture exposure in ceramic industry may cause genotoxic damage in Turkish ceramic industry workers. Similarly, Aksu et al. (2018) were found that occupational exposure to welding fumes may cause genotoxic damage that can lead to important health problems in the workers. In a study with male boron industry workers in Turkey, boron did not induce DNA damage even under extreme exposure conditions (Başaran et al. 2018).

3.4. Nutritional Studies

In recent years, it has been concluded that diets rich in fruit and vegetables are associated with a lower risk of chronic diseases (Ortega 2006).

In a study with 50 smokers and 50 non-smokers, it was seen that supplementation of the diet with 100 mg/day vitamin C, 50 mg/kg vitamin E and 25 mg/kg β-carotene for 20 weeks resulted a decrease in endogenous oxidative base damage in both smokers and non-smokers (Hartmann et al. 1995). In another study with 23 nonsmokers, diet with 330 ml tomato juice, 40 mg lycopene, carrot juice with 22.3 mg β-carotene and 10 g dried spinach power with 11.3 mg lutein reduced endogenous levels of DNA damage in Comet assay (Collins, Dobson, et al. 1997).

Byers and Perry (1992) examined the DNA damage of 3 males and 3 females who had normal breakfast, breakfast with 35 mg/kg bw vitamin C or vitamin C alone by Comet assay. It was determined that comet length of breakfast with vitamin C subjects were lower than the others (Byers and Perry 1992).

According to previous studies, it can be claimed that phytoestrogens can function as antioxidants under certain conditions and protect against oxidatively-induced DNA damage (Sierens et al. 2001).

3.5. Clinical Studies

Comet assay can be used in different clinical studies such as infertility, diabetes, fetal loss, cardiovascular diseases, cancer, etc. The Comet assay has been applied in several clinical studies to investigate the consequences of certain pathological conditions or therapeutical exposure to chemicals at the cellular level (Kassie, Parzefall, and Knasmüller 2000). This assay is for the determination of correlation between the levels of DNA damage and the diseases (Gunasekarana, Raj, and Chand 2015).

The first usage of Comet assay in clinical area was done by Ostling et al. (1987) with applying the neutral version of method to determine the DNA damage of tumor cells from patients receiving radiotherapy for Hodgkin's disease, non-Hodgkin's lymphoma, squamous cell carcinoma or adenocarcinoma.

Collins et al. (1998) concluded that DNA damage of lymphocytes is thus a useful marker of oxidative stress, and in particular Fpg-sensitive sites seem to represent changes specifically related to hyperglycemia due to their study with 10 diabetic patients and their controls. In a study with 63 diabetic patients (15 insulin dependent and 48 non-insulin dependent) and 30 control subjects, significant differences were detected between control and diabetic patients in terms of frequencies of damaged cells and the extend of DNA migration was higher in non-insulin dependent patients by comparison with insulin dependent patients (Şardaş et al. 2001). Pitozzi et al. (2003) used Comet assay to study the oxidative DNA damage in peripheral blood cells of type 2 diabetic patients. According to this study, it can be said that the measurement of oxidative DNA damage in leukocytes by means of the comet assay is a suitable marker for the evaluation of systemic oxidative stress in diabetic patients.

Kan et al. (2002) reported the role of vitamin E supplementation on DNA damage in dialysis patients. Their results demonstrated that the DNA breakage observed in the lymphocytes of patients before vitamin E supplementation was significantly higher than in the controls.

It is already known that there is a relationship between coronary artery disease and DNA damage. The data from the study of Demirbag et al. (2005) showed that the level of DNA damage is increased in 53 coronary artery patients when compared to 42 controls.

The levels of oxidative damage in the peripheral lymphocytes of 24 Alzheimer's disease patients and 21 age-matched controls were determined by Comet assay applied to freshly isolated blood samples with oxidative lesion-specific DNA repair endonucleases (endonuclease III for oxidized pyrimidines, Fpg for oxidized purines) by Kadioglu et al. (2004). It was demonstrated that Alzheimer's disease is associated with elevated levels of oxidized pyrimidines and purines ($p < 0.0001$) as compared with age-matched control subjects.

Portich et al. (2017) detected DNA damage of bone marrow samples of 28 pediatric acute lymphoid leukaemia patients. The results showed that there were no differences during therapy. Lower DNA damage was seen longer time after treatment in comparison at the day of diagnosis. In a study with 146 bladder cancer patients and 141 paired controls, it was found that DNA damage of peripheral blood mononuclear cells (PBMC) were significantly higher than controls (Allione et al. 2018). Cuchra et al. (2016) detected higher oxidative DNA damage and lower DNA repair in breast cancer patients when compared with controls. Similarly, Paz et al. (2018) found higher DNA damage in breast cancer patients in all stages of the treatment. DNA damage of blood samples of 44 colorectal patients was determined. Higher DNA damage was seen in case of malnutrition and after treatment (Vodicka et al. 2019). Sestakova et al. (2016) also examined the DNA damage of lymphocytes in 59 germ cell cancer patients. Higher DNA damage and correlation with presence of metastasis and serum tumour marker levels were seen.

Most of the studies were concluded that there was a relationship between DNA damage of sperm and infertility (Lewis and Agbaje 2008).

Comet assay is found to be a more sensitive technique in the evaluation of sperm DNA damage and fragmentation compared to other methods (Gunasekarana, Raj, and Chand 2015).

CONCLUSION

With the advantages of Comet assay, it is easy to get large amounts of data in human biomonitoring studies in a very short period. However, it is very important to be aware of its limitations. The future usage of Comet assay could impact some other important areas. The Comet assay will be a good addition to the currently existing tests for human biomonitoring studies.

REFERENCES

Aksu, İ., Anlar, H. G., Taner, G., Bacanlı, M., İritaş, S., Tutkun, E. & Basaran, N. (2018). Assessment of DNA damage in welders using comet and micronucleus assays. *Mutation Research/ Genetic Toxicology and Environmental Mutagenesis*, doi: 10.1016/j.mrgentox.2018.11.006.

Allione, A., Pardini, B., Viberti, C., Oderda, M., Allasia, M., Gontero, P., Vineis, P., Sacerdote, C. & Matullo, G. (2018). The prognostic value of basal DNA damage level in peripheral blood lymphocytes of patients affected by bladder cancer. *Urologic Oncology: Seminars and Original Investigations*, 35(5), 241.e15-241.e23.

Anlar, H. G., Taner, G., Bacanli, M., Iritas, S., Kurt, T., Tutkun, E., Yilmaz, O. H. & Basaran, N. (2017). Assessment of DNA damage in ceramic workers. *Mutagenesis*, 33 (1), 97-104.

Anlar, H. G., Bacanli, M., Çal, T., Aydin, S., Ari, N., Ündeğer Bucurgat, Ü., Başaran, A. A. & Başaran, N. (2018). Effects of cinnamic acid on complications of diabetes. *Turkish Journal of Medical Sciences*, 48 (1), 168-177.

Aydın, S., Bacanlı, M., Taner, G., Şahin, T., Başaran, A. A. & Başaran, N. (2013). Protective effects of resveratrol on sepsis-induced DNA damage in the lymphocytes of rats. *Human & Experimental Toxicology*, *32* (10), 1048-1057.

Aydın, S., Bacanlı, M., Anlar, H. G., Çal, T., Arı, N., Ündeğer Bucurgat, Ü., Başaran, A. A. & Başaran, N. (2019). Preventive role of Pycnogenol® against the hyperglycemia-induced oxidative stress and DNA damage in diabetic rats. *Food and Chemical Toxicology*, *124*, 54-63.

Aydın, S., Şahin, T. T., Bacanlı, M., Taner, G., Başaran, A. A., Aydın, S. & Başaran, N. (2016). Resveratrol protects sepsis-induced oxidative DNA damage in liver and kidney of rats. *Balkan Medical Journal*, *33* (6), 594.

Azqueta, A., Langie, S., Boutet-Robinet, E., Duthie, S., Ladeira, C., Møller, P., Collins, A. & Godschalk, R. W. L. (2019). DNA repair as a human biomonitoring tool; comet assay approaches. *Mutation Research/Reviews in Mutation Research*, *781*, 71-87.

Azqueta, A., Langie, S., Slyskova, J. & Collins, A. R. (2013). Measurement of DNA base and nucleotide excision repair activities in mammalian cells and tissues using the comet assay–a methodological overview. *DNA Repair*, *12* (11), 1007-1010.

Azqueta, A., Slyskova, J., Langie, S., O'Neill Gaivão, I. & Collins, A. (2014). Comet assay to measure DNA repair: approach and applications. *Frontiers in Genetics*, *5*, 288.

Bacaksiz, A., Kayaalti, Z., Soylemez, E., Tutkun, E. & Soylemezoglu, T. (2014). Lymphocyte DNA damage in Turkish asphalt workers detected by the comet assay. *International Journal of Environmental Health Research*, *24* (1), 11-17.

Bacanlı, M., Aydın, S., Taner, G., Göktaş, H. G., Şahin, T., Başaran, A. A. & Başaran, N. (2016). Does rosmarinic acid treatment have protective role against sepsis-induced oxidative damage in Wistar Albino rats? *Human & Experimental Toxicology*, 35 (8), 877-886.

Bacanlı, M., Anlar, H. G., Aydın, S., Çal, T., Arı, N., Ündeğer Bucurgat, Ü., Başaran, A. A. & Başaran, N. (2017). d-limonene ameliorates diabetes and its complications in streptozotocin-induced diabetic rats. *Food and Chemical Toxicology*, 110, 434-442.

Bacanlı, M., Aydın, S., Anlar, H. G., Çal, T., Arı, N., Ündeğer Bucurgat, Ü., Başaran, A. A. & Başaran, N. (2018). Can ursolic acid be beneficial against diabetes in rats? *Turkish Journal of Biochemistry*, *43* (5), 520-529.

Bacanli, M., Aydin, S., Anlar, H. G., Çal, T., Ündeğer Bucurgat, Ü., Ari, N., Başaran, A. A. & Başaran, N. (2018). Protective Effects of Ursolic Acid in the Kidneys of Diabetic Rats. *Turkish Journal of Pharmaceutical Sciences*, *15*(2), 166-170.

Bacanlı, M., Aydın, S., Taner, G., Göktaş, H. G., Şahin, T., Başaran, A. A. & Başaran, N. (2014). The protective role of ferulic acid on sepsis-induced oxidative damage in Wistar albino rats. *Environmental toxicology and pharmacology*, *38* (3), 774-782.

Bacanlı, M., Başaran, A. A. & Başaran, N. (2015). The antioxidant and antigenotoxic properties of citrus phenolics limonene and naringin. *Food and chemical Toxicology*, *81*, 160-170.

Bacanlı, M., Başaran, A. A. & Başaran, N. (2017). The antioxidant, cytotoxic, and antigenotoxic effects of galangin, puerarin, and ursolic acid in mammalian cells. *Drug and chemical toxicology*, *40* (3), 256-262.

Başaran, N., Duydu, Y., Üstündağ, A., Taner, G., Aydin, S., Anlar, H. G., Yalçin, C. Ö., Bacanli, M., Aydos, K. & Atabekoğlu, C. S. (2018). Evaluation of the DNA damage in lymphocytes, sperm and buccal cells of workers under environmental and occupational boron exposure conditions. *Mutation Research/Genetic Toxicology and Environmental Mutagenesis*, doi: 10.1016/j.mrgentox.2018.12.013.

Bhalli, J. A., Khan, Q. M. & Nasim, A. (2006). DNA damage in Pakistani pesticide-manufacturing workers assayed using the Comet assay. *Environmental and molecular mutagenesis*, *47* (8), 587-593.

Byers, T. & Perry, G. (1992). Dietary carotenes, vitamin C, and vitamin E as protective antioxidants in human cancers. *Annual review of Nutrition*, *12* (1), 139-159.

Collins, A., Dušinská, M., Franklin, M., Somorovská, M., Petrovská, H., Duthie, S., Fillion, L., Panayiotidis, M., Rašlová, K. & Vaughan, N. (1997). Comet assay in human biomonitoring studies: reliability, validation, and applications. *Environmental and molecular mutagenesis*, *30* (2), 139-146.

Collins, A. R., Dobson, V. L., Dušinská, M., Kennedy, G. & Štětina, R. (1997). The comet assay: what can it really tell us? *Mutation Research/Fundamental and Molecular Mechanisms of Mutagenesis*, *375* (2), 183-193.

Collins, A. R., Rašlová, K., Somorovská, M., Petrovská, H., Ondrušová, A., Vohnout, B., Fábry, R. & Dušinská, M. (1998). DNA Damage in Diabetes: Correlation with a Clinical Marker. *Free Radical Biology and Medicine*, *25* (3), 373-377.

Cortés-Gutiérrez, E. I., López-Fernández, C., Fernández, J. L., Dávila-Rodríguez, M. I., Johnston, S. D. & Gosálvez, J. (2014). Interpreting sperm DNA damage in a diverse range of mammalian sperm by means of the two-tailed comet assay. *Frontiers in genetics*, *5*, 404.

Cuchra, M., Mucha, B., Markiewicz, L., Przybylowska-Sygut, K., Pytel, D., Jeziorski, A., Kordek, R. & Majsterek, I. (2016). The role of base excision repair in pathogenesis of breast cancer in the Polish population. *Molecular carcinogenesis*, *55* (12), 1899-1914.

Dalrymple, A., Ordoñez, P., Thorne, D., Dillon, D. & Meredith, C. (2015). An improved method for the isolation of rat alveolar type II lung cells: Use in the Comet assay to determine DNA damage induced by cigarette smoke. *Regulatory Toxicology and Pharmacology*, *72* (1), 141-149.

Demirbag, R., Yilmaz, R. & Kocyigit, A. (2005). Relationship between DNA damage, total antioxidant capacity and coronary artery disease. *Mutation Research/Fundamental and Molecular Mechanisms of Mutagenesis*, *570* (2), 197-203.

Dobrzyńska, M. M. (2014). DNA damage in organs of female and male mice exposed to nonylphenol, as a single agent or in combination with ionizing irradiation: a comet assay study. *Mutation Research/Genetic Toxicology and Environmental Mutagenesis*, *772*, 14-19.

Everatt, R., Slapšytė, G., Mierauskienė, J., Dedonytė, V. & Bakienė, L. (2013). Biomonitoring study of dry cleaning workers using cytogenetic tests and the comet assay. *Journal of occupational and environmental hygiene*, *10* (11), 609-621.

Gunasekarana, V., Raj, G. V. & Chand, P. (2015). A comprehensive review on clinical applications of comet assay. *Journal of clinical and diagnostic research*, *9* (3), GE01.

Hartmann, A., Nieβ, A. M., Grünert-Fuchs, M., Poch, B. & Speit, G. (1995). Vitamin E prevents exercise-induced DNA damage. *Mutation Research Letters*, *346* (4), 195-202.

Kadioglu, E., Sardas, S., Aslan, S., Isik, E. & Karakaya, A. E. (2004). Detection of oxidative DNA damage in lymphocytes of patients with Alzheimer's disease. *Biomarkers*, *9* (2), 203-209.

Kan, E., Ündeğer, Ü., Bali, M. & Başaran, N. (2002). Assessment of DNA strand breakage by the alkaline COMET assay in dialysis patients and the role of Vitamin E supplementation. *Mutation Research/Genetic Toxicology and Environmental Mutagenesis*, *520* (1), 151-159.

Kassie, F., Parzefall, W. & Knasmüller, S. (2000). Single cell gel electrophoresis assay: a new technique for human biomonitoring studies. *Mutation Research/Reviews in Mutation Research*, *463* (1), 13-31.

Kianmehr, M., Hajavi, J. & Gazeri, J. (2017). Assessment of DNA damage in blood lymphocytes of bakery workers by comet assay. *Toxicology and industrial health*, *33* (9), 726-735.

Kryston, T. B., Georgiev, A. B., Pissis, P. & Georgakilas, A. G. (2011). Role of oxidative stress and DNA damage in human carcinogenesis *Mutation Research/Fundamental and Molecular Mechanisms of Mutagenesis*, *711* (1-2), 193-201.

Langie, S. A. S., Azqueta, A. & Collins, A. R. (2015). The comet assay: past, present, and future. *Frontiers in genetics*, *6*, 266.

Lewis, S. E. M. & Agbaje, I. M. (2008). Using the alkaline comet assay in prognostic tests for male infertility and assisted reproductive technology outcomes. *Mutagenesis*, *23* (3), 163-170.

Megyesi, J., Biró, A., Wigmond, L., Major, J. & Tompa, A. (2014). Use of comet assay for the risk assessment of oil-and chemical-industry workers. *Orvosi hetilap*, *155* (47), 1872-1875.

Ortega, R. M. (2006). Importance of functional foods in the Mediterranean diet. *Public health nutrition*, *9* (8A), 1136-1140.

Ostling, O., Johanson, K. J., Blomquist, E. & Hagelqvist, E. (1987). DNA damage in clinical radiation therapy studied by microelectrophoresis in single tumour cells. A preliminary report. *Acta oncologica (Stockholm, Sweden)*, *26* (1), 45-48.

Paz, M. F. C. J., Alencar, M. V. O. B., Junior, A. L. G., Machado, K. C., Islam, M. T., Ali, E. S., Shill, M. C., Ahmed, M., Uddin, S. J. & Mata, A. M. O. F. (2018). Correlations between Risk Factors for Breast Cancer and Genetic Instability in Cancer Patients—A Clinical Perspective Study. *Frontiers in genetics*, *8*, 236.

Pitozzi, V., Giovannelli, L., Bardini, G., Rotella, C. M. & Dolara, P. (2003). Oxidative DNA damage in peripheral blood cells in type 2 diabetes mellitus: higher vulnerability of polymorphonuclear leukocytes. *Mutation Research/Fundamental and Molecular Mechanisms of Mutagenesis*, *529* (1), 129-133.

Portich, J. P., Santos, R. P., Kersting, N., Jorge, K. B., Casagrande, P. R., Costa, G. S., Cionek, J. M. G. D., Olguins, D. F., Sinigaglia, M. & Busatto, F. F. (2017). DNA damage response in patients with pediatric Acute Lymphoid Leukemia during induction therapy. *Leukemia research*, *54*, 59-65.

Ramos, A. A., Azqueta, A., Pereira-Wilson, C. & Collins, A. R. (2010). Polyphenolic compounds from Salvia species protect cellular DNA from oxidation and stimulate DNA repair in cultured human cells. *Journal of Agricultural and Food Chemistry*, *58* (12), 7465-7471.

Rojas, E., Lopez, M. C. & Valverde, M. (1999). Single cell gel electrophoresis assay: methodology and applications. *Journal of Chromatography B: Biomedical Sciences and Applications*, *722* (1-2), 225-254.

Rojas, E., Lorenzo, Y., Haug, K., Nicolaissen, B. & Valverde, M. (2014). Epithelial cells as alternative human biomatrices for comet assay. *Frontiers in genetics*, 5, 386.

Sestakova, Z., Kalavska, K., Hurbanova, L., Jurkovicova, D., Gursky, J., Chovanec, M., Svetlovska, D., Miskovska, V., Obertova, J. & Palacka, P. (2016). The prognostic value of DNA damage level in peripheral blood lymphocytes of chemotherapy-naïve patients with germ cell cancer. *Oncotarget*, 7 (46), 75996.

Shah, A. J., Lakkad, B. C. & Rao, M. V. (2016). Genotoxicity in lead treated human lymphocytes evaluated by micronucleus and comet assays. *Indian Journal of Experimental Biology*, 54(8), 502-508.

Sierens, J., Hartley, J. A., Campbell, M. J., Leathem, A. J. C. & Woodside, J. V. (2001). Effect of phytoestrogen and antioxidant supplementation on oxidative DNA damage assessed using the comet assay. *Mutation Research/DNA Repair*, 485 (2), 169-176.

Silva, J. P., Gomes, A. C. & Coutinho, O. P. (2008). Oxidative DNA damage protection and repair by polyphenolic compounds in PC12 cells. *European Journal of Pharmacology*, 601 (1-3), 50-60.

Singh, Z. & Chadha, P. (2014). DNA damage due to inhalation of complex metal particulates among foundry workers. *Advances in Environmental Biology*, 8 (15), 225-230.

Singh, Z. (2018). Comet assay as a sensitive technique in occupational health studies; A literature review. *Journal of Occupational Health and Epidemiology*, 7 (4), 240-245.

Şardaş, S., Yilmaz, M., Öztok, U., Çakir, N. & Karakaya, A. E. (2001). Assessment of DNA strand breakage by comet assay in diabetic patients and the role of antioxidant supplementation. *Mutation Research/Genetic Toxicology and Environmental Mutagenesis*, 490 (2), 123-129.

Taner, G., Aydın, S., Bacanlı, M., Sarıgöl, Z., Şahin, T., Başaran, A. A. & Başaran, N. (2014). Modulating effects of pycnogenol® on oxidative stress and DNA damage induced by sepsis in rats. *Phytotherapy research*, 28 (11), 1692-1700.

Tokac, M., Bacanli, M., Dumlu, E. G., Aydın, S., Engin, M., Bozkurt, B., Yalçın, A., Erel, Ö., Kılıç, M. & Başaran, N. (2017). The Ameliorative Effects of Pycnogenol® on Liver Ischemia-Reperfusion Injury in Rats. *Turkish Journal of Pharmaceutical Sciences, 14* (3), 257-263.

Vodicka, P., Vodenkova, S., Opattova, A. & Vodickova, L. (2019). DNA damage and repair measured by comet assay in cancer patients. *Mutation Research/Genetic Toxicology and Environmental Mutagenesis* doi:10.1016/j.mrgentox.2019.05.009.

Zhang, T., Zhao, Q., Zhang, Y. & Ning, J. (2014). Assessment of genotoxic effects of flumorph by the comet assay in mice organs. *Human & experimental toxicology, 33* (3), 224-229.

In: A Closer Look at the Comet Assay
Editor: Keith H. Harmon
ISBN: 978-1-53611-028-9
© 2019 Nova Science Publishers, Inc.

Chapter 2

COMET ASSAY IN OCCUPATIONAL TOXICOLOGY STUDIES

Hatice Gül Anlar[*], *PhD*
Department of Pharmaceutical Toxicology, Faculty of Pharmacy,
Zonguldak Bulent Ecevit University, Zonguldak, Turkey

ABSTRACT

The comet assay, also known as single cell gel electrophoresis (SCGE), is a gel electrophoresis method used to visualize and measure DNA strand breaks in individual cells, using microscopy. It has been found to be a very sensitive, rapid, reliable and fairly inexpensive way of measuring DNA damage. It has a further advantage that the observations are made at the single cell level. It has been increasingly used in occupational biomonitoring studies with different occupational settings. Isolated lymphocytes were the most used cell line in these studies but also exfoliated cells like a nasal and buccal cell, liver, kidney and sperm cells can be used. For example, according to our previous studies with ceramic workers and welders, the results showed that these workers had significantly more DNA damage in their isolated lymphocytes and whole blood compared to controls and also, DNA damage in lymphocytes and whole blood cells were correlated. It was important because using whole

---

[*] Corresponding Author's Email: haticegulanlar@gmail.com.

blood is simpler since the isolation of lymphocytes requires more time and chemicals. This chapter aims to provide knowledge about Comet assay and its usage in occupational toxicology studies.

Keywords: comet assay, occupational toxicology, workers, health, genotoxicity

INTRODUCTION

Most cancers result from industrial and environmental exposures such as occupational tobacco smoke, chemical pollutants in the air, water and food, drugs, radiation, and dietary constituents. It has been estimated that without these environmental and industrial factors, cancer incidence would be dramatically reduced, by as much as 80–90% (Perera and Dickey 1997). Since a person spends, approximately, one/third of his/her life at his/her workplace, the environment where he/she works can be a major factor in evaluating health status. It is estimated that about 4% of all cancers are due to the occupation, among these lung and bladder cancers are strongly dependent on occupational exposure (Doll and Peto 1981; Field and Withers 2012; Silverman et al. 1989). Occupational cancer can largely be prevented through reformed working conditions and practices. Recently, developed countries have taken precaution to control many occupational exposures and the proportion of occupational cancers is therefore likely to fall in these countries. But in developing countries, increasing industrialization brings with it new hazardous chemicals, is often inadequately regulated. There are several factors to classify a country as developed or developing countries, but the main criteria are Gross National Product (GNP) per capita according to income. Countries are classified into three categories i.e., low middle, and high-income. Low and middle-income countries are generally recognized as developing countries and apprenticeship in small workshops is very common and also government regulations are not enough to correspond to new scientific data in these countries (LaDou 1996).

This chapter aims to provide knowledge about comet assay and its usage in occupational toxicology studies in various occupational groups in different countries.

Comet Assay (Single Cell Gel Electrophoresis (SCGE) Assay)

The comet assay, also known as single cell gel electrophoresis (SCGE), is an electrophoretic technique for direct imaging of DNA damage in individual cells. This method is applicable to *in vitro*, *in vivo*, and *ex vivo* systems that can detect DNA fractures. This assay has quite a simple methodology and high sensitivity. Meanwhile, it requires small numbers of cells to rapid production of data which are the important advantages of this assay over other genotoxicity tests (Collins 2004).

For the first time, Rydberg and Johanson (Rydberg and Johanson 1978) quantitated directly the levels of DNA damage in individual cells after embedding them in agarose gel on slides and lysing under mild alkali conditions to allow for the partial unwinding of DNA. Then the cells were neutralized and stained with acridine orange. The level of DNA damage was measured by calculating the ratio of green (indicating double-stranded DNA) to red (indicating single-stranded DNA) fluorescence using a photometer. In order to enhance the sensitivity of detecting DNA damage in single cells, Ostling and Johanson (Ostling and Johanson 1984) used the microgel electrophoresis technique to determine DNA damage. In this method, single and double strand fractures in the DNA chain, alkaline volatile regions, and oxidative DNA base damage can be determined.

In this method, a small number of cells are attached to the microscope slide in agar gel. Cells are lysed and the DNA is opened at different pHs. Selecting different pHs for electrophoresis allows to measure different degrees of damage. In this assay, cells can be electrophoresed under neutral or alkali conditions. Using neutral conditions for the lysis and electrophoresis allows the detection of double-strand breaks but it cannot detect single strand ones. A lot of agents induce 5- to 2000-fold more single strand breaks than double-strand breaks, the use of neutral conditions is not as sensitive as alkaline conditions in analyzing the DNA damage. Singh and colleagues (N. P. Singh et al. 1988) developed a modified method which is capable of detecting DNA single-strand breaks under alkaline conditions in 1988.

The degree of DNA migration gives information about the extent of DNA damage in the cell (Cemeli, Baumgartner, and Anderson 2009). Since the undamaged DNA is large, it migrated without a tail while the damaged DNA disintegrates into small fragments during the application of current in the electrophoresis (McArt et al. 2009). Afterward, these images, called COMET, are measured and evaluated. Ethidium bromide (EtBr) is the most preferred dye for the staining. It allows counting in a semi-dark environment and is well connected to double chain fractures due to its bright fluorescent color and this light does not disappear quickly. EtBr also causes a very low background signal. 4,6-diamidino-2-phenylindole (DAPI), 1,1'-(4,4,8,8-tetramethyl-4,8-diazaundecamethylene)bis[4-[(3-methyl benzo-1,3-oxazole-2-yl)methylidene]-1,4 dihydroquinolinium] tetraiodide (YOYO), propidium iodide and SYBR® are other fluorescent dyes using the comet assay. YOYO is an expensive and long-lasting dye which is used in cells that include low DNA and it gives very strong fluorescent light. But it is expensive and can not be stored for a long time. SYBR® also gives very strong fluorescent light, but this can be reduced during counting. Studies have shown that the effectiveness of EtBr does not differ from these dyes. The silver dyeing is an inexpensive method that allows counting without the need for a fluorescent microscope. However, the counting takes a long time and the background signal may be excessive (Tice et al. 2000; Gichner, Mukherjee, and Velemínský 2006).

In the COMET assay, % DNA in the tail, % DNA in the head, Olive tail moment, tail length, tail density, etc. values can be used. Tail length data has been criticized because it is only linear with respect to dose within a narrow interval. However, when calculating the Olive tail moment, the length between the center of the head and tail must also be known (μm). Therefore, the imaging system must be calibrated before counting, otherwise, it is not possible to make comparisons between laboratories. On the other hand, in most studies which this value was given, the length unit was not shown. In addition, this value may vary depending on the electrophoresis conditions and how the center of mass is defined in the program. Therefore, in evaluating and interpreting the results, it is recommended that the concentration of the DNA fragment in the tail (% tail intensity) should be used to express the percentage value of the whole cell according to the density of the tail (Kumaravel and Jha 2006; Kumaravel et al. 2009). Many computer programs are available to perform the evaluation. The degree of determined DNA damage depends on many physical and technical factors. Lysis conditions (salt concentration, pH, lysis time, etc.) may affect the results. Salt residues in the lysis solution can inhibit DNA migration, and the absence of salt may require lower electrophoresis voltage (Olive et al. 1992). Increasing agar concentration and decreasing electrophoresis time increase the probability of DNA damage detection (Olive, Banath, and Durand 1990; Vijayalaxmi, Tice, and Strauss 1992).

The COMET method allows working on different cell types while lymphocytes are the most preferred cells because they can be taken easily and give good results (Faust et al. 2004). Detection of DNA damage in lymphocytes may be affected the age, physical activity, and smoking status of the person. But it has some disadvantages. Firstly blood cells are not actual target tissue for various cancers, so it is not clear that the damage detected in white blood cells reflects the damage in target tissues. Sometimes human tissue removed at surgery can be investigated by comet assay but the necessary control tissue from healthy individuals is really hard to obtain. Besides lymphocytes, sperm, cheek and nasal epithelium and placental cells can also be used in this method (Narendra P. Singh et al. 1989; Fairbairn, Olive, and O'Neill 1995). But it seems that the cytostructure of epithelial cells inhibit the release of DNA, and extensive lysis and digestion with proteases are necessary (Collins 2004).

Although alkaline comet assay has been widely used for many years, method validation studies have started recently. As a result of the studies carried out by the Japan Alternative Methods Validation Center (JaCVAM) and the European Alternative Methods Validation Center (ECVAM), *in vivo* comet assay in mammalian cells was validated by the Organization of

Economic Development and Cooperation (OECD) on September 26, 2014 (Test number: 489) (Uno et al. 2015; OECD 2016).

Occupational Toxicology Studies with the Comet Assay

Comet assay; which investigates the exposure of workers in the ceramic industry, pottery, welders, plastics industry, lamination plants, footwear factories, cigarette factories, and pharmaceutical production; has been used in many studies. Most of these studies showed increased DNA damage due to occupational exposure and some of these studies provided corroborating data from multiple biomarkers. But, in some studies, significant DNA damage was not detected by comet assay associated with occupational exposure in the workplace such as biomonitoring studies of workers exposed to boron (Duydu et al. 2012), formaldehyde (Aydın et al. 2013), sewage (Friis et al. 1997), butadiene (Sram et al. 1998; Tates et al. 1996), 4,40-methylenediphenyldiisocyanate, 2,4-toluene diisocyanate and 2,6-toluene diisocyanate (Marczynski et al. 2005), radiofrequency radiation (Maes, Van Gorp, and Verschaeve 2006) and air pollutants from traffic fumes (Carere et al. 2002), hair dyes (Şardaş, Aygün, and Karakaya 1997) and also, waste disposal workers (Hartmann, Fender, and Speit 1998) shoe factory workers (Pitarque et al. 1999) coke oven (van Delft et al. 2001; Siwinska, Mielzynska, and Kapka 2004) and rubber factory workers (Moretti et al. 1996), flight personnel exposed to cosmic radiation and environmental pollutants (Cavallo et al. 2002).

In our laboratory, the comet assay has been widely used for a long time in different occupational groups. Also, we performed comet assay in animals and cell cultures to evaluate the genotoxic and anti-genotoxic effects of natural phenolic compounds. A first occupational study in our laboratory using comet assay was conducted in professional nurses employed in the oncology departments. Chemotherapic agents have been widely used in the treatment of cancer but careless handling of these drugs may lead to exposure and mutagenic effects. Health care professionals who involved in the preparation and administration of cytotoxic agents should be educated about the possible hazards. In this study, blood samples were collected from 30 professional nurses employed in the oncology departments for at least 6 months and 30 controls with comparable to nurses regarding the age, sex and smoking habits, not practicing in the chemotherapy services. Lymphocytes were isolated and DNA damage was examined by the alkaline single cell gel electrophoresis. Work characteristics of the exposed nurses and the use of personal protective equipment were also investigated. The DNA damage in the lymphocytes of the nurses was significantly higher than the controls. DNA damage of the nurses applying the necessary individual safety protections during their work was found to be significantly lower than other nurses. Cigarette smoking and duration of exposure to antineoplastic drugs did not significantly affected DNA damage (Undeger et al. 1999).

Ionizing radiation has been commonly used in the treatment of cancer but exposure to ionizing radiation is a great concern in the staff who are at risk of such occupational exposure. In a study with 30 technicians employed in radiation oncology departments for at least 1 year and an equal number of controls who were not exposed to radiation or chemotherapy. It was shown that DNA damage in the isolated lymphocytes of these technicians was significantly higher than that in the controls. There was a significant association between DNA damage and cigarette smoking and also DNA damage and duration of exposure (Undeger, Zorlu, and Basaran 1999).

Pesticides are another mutagenic agent which is commonly used in domestic and industrial applications. Various studies have revealed a significantly elevated risk for particular tumors in humans exposed to some pesticides. In this study DNA damage were analyzed by comet assay in peripheral lymphocytes of 33 pesticide-exposed workers employed in the municipality of Ankara (Turkey) for at least 1 year of exposure and compared with 33 controls which were not occupationally exposed to pesticides. The DNA damage observed in lymphocytes of the workers was found to be significantly higher than that in the controls. The observed DNA damage was found to be significantly lower in workers who are applying some of the necessary individual safety protections during their work. There were any significant association cigarette smoking, duration of exposure and the degree of DNA damage (Undeger, Zorlu, and Basaran 1999).

Foundry and pottery workers are exposed to a mixture of chemicals and silica is the most important exposure in these workers because it is suspected to cause genetic alterations. In a study of Basaran et al. (Basaran et al. 2003) 30 foundry and 22 pottery workers who are exposed to silica, DNA damage in the peripheral lymphocytes was analyzed by comet assay and compared to 52 healthy subjects with no history of occupational silica or other chemical exposure. These workers had significantly higher DNA damage in lymphocytes when compared to the controls. In workers, it was found that smokers had significantly higher DNA damage than non-smokers.

Recently, boric acid and sodium borates have been classified as being "toxic to reproduction and development" by the European Union, following results of animal studies with high doses. Duydu et. al. (Duydu et al. 2012) investigated the DNA damage of male workers in the boric acid/borate production zone at Bandırma, Turkey. Two versions of the comet assay, the neutral and alkaline, were conducted in sperm cells and there was no statistically significant difference between workers and controls. When the relationship between DNA-strand breaks and motility/morphology parameters of sperm samples were analyzed, there were weak but statistically significant correlations in neutral comet assay but statistically not significant correlation in the alkaline comet assay.

Formaldehyde (FA) is an important chemical and widely used in various application in industry. It has been classified as carcinogenic to humans by International Research on Cancer (IARC) and occupational toxicology limits values were settled. But especially in the developing countries exposure to FA is a still health concern. In a study of Aydin et al. (Aydin et al. 2013) DNA damage in peripheral lymphocytes of workers in medium density fiberboard plants exposed to FA was evaluated by comet assay. They did not find increased DNA damage in the workers compared to controls.

In my study with ceramic workers, blood samples were collected from 99 male ceramic workers and 81 male controls who are comparable to workers regarding age and life habits. Ceramic workers exposed to a complex mixture of chemicals such as silica, inorganic lead, lime, beryllium, and aluminum that can be associated with an increased risk of several diseases. After the blood collection, lymphocytes were isolated and comet assay was done with lymphocytes and also whole blood cells. The results of this study, DNA damage in whole blood and isolated lymphocytes of workers was significantly higher than the controls. Workers older than 42 years and workers working in the ceramic plant more than 16 years have significantly higher DNA damage in the lymphocytes and the whole blood when compared to the other workers which indicate that accumulation of exposure was detected by comet assay. But surprisingly, there was no statistically significant correlation between DNA damage and alcohol usage, smoking, and usage of protective measures. Workers with silicosis have slightly higher DNA damage compared to the other workers. This might show that DNA damage which was detected by comet assay could be an indicator of silicosis susceptibility. Also, there was a positive significant correlation between DNA damage levels in lymphocytes and whole blood. This is really important for the application of the comet assay, since the isolation of lymphocytes is required more time and chemicals while using whole blood is very easy and allow to large biomonitoring studies with comet assay (Anlar et al. 2018).

In another study of our research group, we investigated DNA damage in welders who are exposed to a number of hazardous compounds such as ultraviolet (UV) radiation, electromagnetic fields, toxic metals, and polycyclic aromatic hydrocarbons (PAHs) in addition to welding fumes. In this study, 48 welders and an equal number of control subjects were evaluated for DNA damage in their whole blood and isolated lymphocytes using the comet assay. DNA damage in the lymphocytes and the whole blood of the workers were found to be significantly higher than the control group. No statistically significant correlation was found between DNA damage and the duration of exposure, age, smoking, alcohol usage of the workers. The effects of protective measures on the DNA damage were not evaluated because the large majority of the workers claimed that they were using the protective measures. Similar to the study with ceramic workers, there was a significant correlation between DNA damage in whole blood and lymphocytes (Aksu et al. 2018).

Another study of boron workers was conducted in two districts of Balıkesir; Bandırma and Bigadic, Turkey. 102 male workers who were occupationally exposed to boron from Bandırma and 110 workers who were occupationally and environmentally exposed to boron from Bigadic participated in this study. DNA damage in the sperm and blood cells of 212 males was evaluated by comet assay. No significant increase in the DNA damage in blood and sperm was observed. Also, no correlations were seen between blood boron levels and tail intensity values of the sperm and lymphocyte samples (Başaran et al. 2018).

Sardaş and colleagues (Sardas et al. 1998) collected blood samples from 66 operating room personnel who were exposed to various anesthetics such as halothane, nitrous oxide, and isoflurane. A significant increase in the number of lymphocytes with DNA migration was observed in operating room personnel as compared to controls. Also, smokers had significantly higher DNA damage than nonsmokers. In another study of Sardaş et al. (Sardas et al. 2010) DNA damage was determined in Turkish construction-site workers occupationally exposed to welding fumes (n = 26) by comet assay and they found significantly higher DNA damage in the lymphocytes of workers. They found enhanced damage in smokers compared to nonsmokers, but they did not found a significant correlation between duration of exposure and the DNA damage.

Botta et al. (Botta et al. 2006) showed that French welders (n = 30) had significantly higher DNA damage when compared to the controls. No relationship was found between DNA damage and smoking for either the controls or the welders.

In China, it was observed that welders who had been working in a bus manufacturing industry had significantly higher DNA damage (Zhu, Lam, and Jiang 2001). Sellappa et al. (Sellappa, Prathyumnan, and Balachandar 2010) found that construction workers (n = 96) showed a significant increase in comet tail length compared to controls with adjustment for smoking habits, tobacco chewing, alcohol consumption and years of exposure. It also indicated that chronic occupational exposure to cement during construction work could lead to increased levels of DNA damage and repair inhibition.

Unwinding time and the electrophoresis conditions such as voltage, amperage, and duration are the major technical variables which can affect the sensitivity of the comet assay. In most of the studies with lymphocytes, 20 min for unwinding and 20 min for electrophoresis were performed. The same conditions were applied for buccal, nasal, sperm, and tear samples, while the studies with mononuclear cells, 40 min for unwinding was commonly implemented. It indicated that increasing the unwinding time more than 40 min can cause an increase in comet formation on control slides (Green et al. 1996). On the other hand, Albertini et al. (Albertini et al. 2000) suggested 60 min of unwinding provide maximum sensitivity (Valverde and Rojas 2009).

CONCLUSION

Early identification of hazards is crucial to reduce exposure and carcinogenic risk and the workplace is the main factor which affects the person's health status. Due to these considerations and the complex multi-composition of the workplace environment, the field of biomonitoring has gained interest of scientists and organizations around the world. In this chapter, information about comet assay and occupational toxicology studies with this assay in different occupational settings was given. The most important aim of these studies was the determination of the differences between exposed and control groups with respect to lifestyle variables such as diet, smoking habits, medical treatments, history of chronic diseases, physical activity, etc. But the results of these studies regarding the lifestyle variables, especially the effect of cigarette smoking on DNA damage was contradictory. Therefore, exposure time, a number of cigarettes smoked per day, the kind of tobacco smoked as a confounding factor should be well considered.

The Comet assay has the potential to be applied to all of the categories of human biomonitoring and represents a valuable tool for acquiring knowledge about current levels of exposure to occupational hazards. This assay can be used to monitor large populations of people which is an important advantage of it, but there are also some shortcomings. Therefore, there is a need for future interlaboratory studies to widen the usage of comet assay and to eliminate its disadvantages.

References

Aksu, İ., H. G. Anlar, G. Taner, M. Bacanlı, S. İritaş, E. Tutkun, and N. Basaran. 2018. "Assessment of DNA damage in welders using comet and micronucleus assays." *Mutation Research/Genetic Toxicology and Environmental Mutagenesis.* https://doi.org/https://doi.org/10.1016/j.mrgentox.2018.11.006. http://www.sciencedirect.com/science/article/pii/S138357181830233X.

Albertini, R. J., D. Anderson, G. R. Douglas, L. Hagmar, K. Hemminki, F. Merlo, A. T. Natarajan, H. Norppa, D. E. Shuker, R. Tice, M. D. Waters, and A. Aitio. 2000. "IPCS guidelines for the monitoring of genotoxic effects of carcinogens in humans. International Programme on Chemical Safety." *Mutat Res* 463 (2): 111-72.

Anlar, H. G., G. Taner, M. Bacanli, S. Iritas, T. Kurt, E. Tutkun, O. H. Yilmaz, and N. Basaran. 2018. "Assessment of DNA damage in ceramic workers." *Mutagenesis* 33 (1): 97-104. https://doi.org/10.1093/mutage/gex016.

Aydin, S., H. Canpinar, U. Undeger, D. Guc, M. Colakoglu, A. Kars, and N. Basaran. 2013. "Assessment of immunotoxicity and genotoxicity in workers exposed to low concentrations of formaldehyde." *Arch Toxicol* 87 (1): 145-53. https://doi.org/10.1007/s00204-012-0961-9.

Basaran, N., M. Shubair, U. Undeger, and A. Kars. 2003. "Monitoring of DNA damage in foundry and pottery workers exposed to silica by the alkaline comet assay." *Am J Ind Med* 43 (6): 602-10. https://doi.org/10.1002/ajim.10222.

Başaran, N. Y. Duydu, A. Üstündağ, G. Taner, S. Aydin, H. G. Anlar, C. Ö. Yalçın, M. Bacanli, K. Aydos, C. S. Atabekoğlu, K. Golka, K. Ickstadt, T. Schwerdtle, M. Werner, S. Meyer, and H. M. Bolt. 2018. "Evaluation of the DNA damage in lymphocytes, sperm and buccal cells of workers under environmental and occupational boron exposure conditions." *Mutation Research/Genetic Toxicology and Environmental Mutagenesis.* https://doi.org/10.1016/j.mrgentox.2018.12.013. http://www.sciencedirect.com/science/article/pii/S1383571818302316.

Botta, C., G. Iarmarcovai, F. Chaspoul, I. Sari-Minodier, J. Pompili, T. Orsiere, J. L. Berge-Lefranc, A. Botta, P. Gallice, and M. De Meo. 2006. "Assessment of occupational exposure to welding fumes by inductively coupled plasma-mass spectroscopy and by the alkaline Comet assay." *Environ Mol Mutagen* 47 (4): 284-95. https://doi.org/10.1002/em.20205.

Carere, A., C. Andreoli, R. Galati, P. Leopardi, F. Marcon, M. V. Rosati, S. Rossi, F. Tomei, A. Verdina, A. Zijno, and R. Crebelli. 2002. "Biomonitoring of exposure to urban air pollutants: analysis of sister chromatid exchanges and DNA lesions in peripheral lymphocytes of traffic policemen." *Mutat Res* 518 (2): 215-24.

Cavallo, D., P. Tomao, A. Marinaccio, B. Perniconi, A. Setini, S. Palmi, and S. Iavicoli. 2002. "Evaluation of DNA damage in flight personnel by Comet assay." *Mutat Res* 516 (1-2): 148-52.

Cemeli, E., A. Baumgartner, and D. Anderson. 2009. "Antioxidants and the Comet assay." *Mutat Res* 681 (1): 51-67. https://doi.org/10.1016/j.mrrev.2008.05.002.

Collins, A. R. 2004. "The comet assay for DNA damage and repair: principles, applications, and limitations." *Mol Biotechnol* 26 (3): 249-61. https://doi.org/10.1385/mb:26:3:249.

Doll, R., and R. Peto. 1981. "The causes of cancer: quantitative estimates of avoidable risks of cancer in the United States today." *J Natl Cancer Inst* 66 (6): 1191-308.

Duydu, Y., N. Basaran, A. Ustundag, S. Aydin, U. Undeger, O. Y. Ataman, K. Aydos, Y. Duker, K. Ickstadt, B. S. Waltrup, K. Golka, and H. M. Bolt. 2012. "Assessment of DNA integrity (COMET assay) in sperm cells of boron-exposed workers." *Arch Toxicol* 86 (1): 27-35. https://doi.org/10.1007/s00204-011-0743-9.

Fairbairn, D. W., P. L. Olive, and K. L. O'Neill. 1995. "The comet assay: a comprehensive review." *Mutat Res* 339 (1): 37-59.

Faust, F., F. Kassie, S. Knasmuller, R. H. Boedecker, M. Mann, and V. Mersch-Sundermann. 2004. "The use of the alkaline comet assay with lymphocytes in human biomonitoring studies." *Mutat Res* 566 (3): 209-29. https://doi.org/10.1016/j.mrrev.2003.09.007.

Field, R. William, and Brian L. Withers. 2012. "Occupational and environmental causes of lung cancer." *Clinics in chest medicine* 33 (4): 681-703. https://doi.org/10.1016/j.ccm.2012.07.001. https://www.ncbi.nlm.nih.gov/pubmed/23153609. https://www.ncbi.nlm.nih.gov/pmc/articles/PMC3875302/.

Friis, L., H. Vaghef, C. Edling, and B. Hellman. 1997. "No increased DNA damage in peripheral lymphocytes of sewage workers as evaluated by alkaline single cell gel electrophoresis." *Occupational and environmental medicine* 54 (7): 494-498. https://doi.org/10.1136/oem.54.7.494. https://www.ncbi.nlm.nih.gov/pubmed/9282126. https://www.ncbi.nlm.nih.gov/pmc/articles/PMC1128819/.

Gichner, Tomás, Anita Mukherjee, and Jirí Velemínský. 2006. *DNA staining with the fluorochromes EtBr, DAPI and YOYO-1 in the comet assay with tobacco plants after treatment with ethyl methanesulphonate, hyperthermia and DNase-I.* Vol. 605.

Green, M. H., J. E. Lowe, C. A. Delaney, and I. C. Green. 1996. "Comet assay to detect nitric oxide-dependent DNA damage in mammalian cells." *Methods Enzymol* 269: 243-66.

Hartmann, A., H. Fender, and G. Speit. 1998. "Comparative biomonitoring study of workers at a waste disposal site using cytogenetic tests and the comet (single-cell gel) assay." *Environ Mol Mutagen* 32 (1): 17-24.

Kumaravel, T. S., and A. N. Jha. 2006. "Reliable Comet assay measurements for detecting DNA damage induced by ionising radiation and chemicals." *Mutat Res* 605 (1-2): 7-16. https://doi.org/10.1016/j.mrgentox.2006.03.002.

Kumaravel, T. S., B. Vilhar, S. P. Faux, and A. N. Jha. 2009. "Comet Assay measurements: a perspective." *Cell Biol Toxicol* 25 (1): 53-64. https://doi.org/10.1007/s10565-007-9043-9.

LaDou, Joseph. 1996. "The role of multinational corporations in providing occupational health and safety in developing countries." *International Archives of Occupational and Environmental Health* 68 (6): 363-366. https://doi.org/10.1007/bf00377852. https://doi.org/10.1007/BF00377852.

Maes, A., U. Van Gorp, and L. Verschaeve. 2006. "Cytogenetic investigation of subjects professionally exposed to radiofrequency radiation." *Mutagenesis* 21 (2): 139-42. https://doi.org/10.1093/mutage/gel008.

Marczynski, B., R. Merget, T. Mensing, S. Rabstein, M. Kappler, A. Bracht, M. G. Haufs, H. U. Kafferlein, and T. Bruning. 2005. "DNA strand breaks in the lymphocytes of workers exposed to diisocyanates: indications of individual differences in susceptibility after low-dose and short-term exposure." *Arch Toxicol* 79 (6): 355-62. https://doi.org/10.1007/s00204-004-0639-z.

McArt, D. G., G. McKerr, C. V. Howard, K. Saetzler, and G. R. Wasson. 2009. "Modelling the comet assay." *Biochem Soc Trans* 37 (Pt 4): 914-7. https://doi.org/10.1042/bst0370914.

Moretti, M., M. Villarini, G. Scassellati-Sforzolini, S. Monarca, M. Libraro, C. Fatigoni, F. Donato, C. Leonardis, and L. Perego. 1996. "Biological monitoring of genotoxic hazard in workers of the rubber industry." *Environ Health Perspect* 104 Suppl 3: 543-5. https://doi.org/10.1289/ehp.96104s3543.

OECD. 2016. "Test No. 489: *In Vivo* Mammalian Alkaline Comet Assay."

Olive, P. L., J. P. Banath, and R. E. Durand. 1990. "Heterogeneity in radiation-induced DNA damage and repair in tumor and normal cells measured using the "comet" assay." *Radiat Res* 122 (1): 86-94.

Olive, P. L., D. Wlodek, R. E. Durand, and J. P. Banáth. 1992. "Factors influencing DNA migration from individual cells subjected to gel electrophoresis." *Experimental Cell Research* 198 (2): 259-267. https://doi.org/https://doi.org/10.1016/0014-4827(92)90378-L. http://www.sciencedirect.com/science/article/pii/001448279290378L.

Ostling, O., and K. J. Johanson. 1984. "Microelectrophoretic study of radiation-induced DNA damages in individual mammalian cells." *Biochem Biophys Res Commun* 123 (1): 291-8.

Perera, F. P., and C. Dickey. 1997. "Molecular epidemiology and occupational health." *Ann N Y Acad Sci* 837: 353-9.

Pitarque, M., A. Vaglenov, M. Nosko, A. Hirvonen, H. Norppa, A. Creus, and R. Marcos. 1999. "Evaluation of DNA damage by the Comet assay in shoe workers exposed to toluene and other organic solvents". *Mutat Res* 441(1):115-27.

Rydberg, B., and K. J. Johanson. 1978. "Estimation of DNA strand breaks in single mammalian cells." In *DNA Repair Mechanisms*, edited by Philip C. Hanawalt, Errol C. Friedberg and C. Fred Fox, 465-468. Academic Press.

Sardas, S., N. Aygun, M. Gamli, Y. Unal, N. Unal, N. Berk, and A. E. Karakaya. 1998. "Use of alkaline comet assay (single cell gel electrophoresis technique) to detect DNA damages in lymphocytes of operating room personnel occupationally exposed to anaesthetic gases." *Mutat Res* 418 (2-3): 93-100.

Sardas, S., G. Z. Omurtag, A. Tozan, H. Gul, and D. Beyoglu. 2010. "Evaluation of DNA damage in construction-site workers occupationally exposed to welding fumes and solvent-based paints in Turkey." *Toxicol Ind Health* 26 (9): 601-8. https://doi.org/10.1177/0748233710374463.

Sellappa, S., S. Prathyumnan, and V. Balachandar. 2010. "DNA damage induction and repair inhibition among building construction workers in South India." *Asian Pac J Cancer Prev* 11 (4): 875-80.

Silverman, D. T., L. I. Levin, R. N. Hoover, and P. Hartge. 1989. "Occupational risks of bladder cancer in the United States: I. White men." *J Natl Cancer Inst* 81 (19): 1472-80. https://doi.org/10.1093/jnci/81.19.1472.

Singh, N. P., M. T. McCoy, R. R. Tice, and E. L. Schneider. 1988. "A simple technique for quantitation of low levels of DNA damage in individual cells." *Exp Cell Res* 175 (1): 184-91.

Singh, N. P., D. B. Danner, R. R. Tice, M. T. McCoy, G. D. Collins, and E. L. Schneider. 1989. "Abundant alkali-sensitive sites in DNA of human and mouse sperm." *Experimental Cell Research* 184 (2): 461-470. https://doi.org/https://doi.org/10.1016/0014-4827(89)90344-3. http://www.sciencedirect.com/science/article/pii/0014482789903443.

Siwinska, E., D. Mielzynska, and L. Kapka. 2004. "Association between urinary 1-hydroxypyrene and genotoxic effects in coke oven workers." *Occup Environ Med* 61 (3): e10. https://doi.org/10.1136/oem.2002.006643.

Sram, R. J., P. Rossner, K. Peltonen, K. Podrazilova, G. Mrackova, N. A. Demopoulos, G. Stephanou, D. Vlachodimitropoulos, F. Darroudi, and A. D. Tates. 1998. "Chromosomal aberrations, sister-chromatid exchanges, cells with high frequency of SCE, micronuclei and comet assay parameters in 1, 3-butadiene-exposed workers." *Mutat Res* 419 (1-3): 145-54.

Şardaş, S, N Aygün, and AE Karakaya. 1997. "Genotoxicity studies on professional hair colorists exposed to oxidation hair dyes." *Mutation Research/Genetic Toxicology and Environmental Mutagenesis* 394 (1-3): 153-161.

Tates, A. D., F. J. van Dam, F. A. de Zwart, F. Darroudi, A. T. Natarajan, P. Rössner, K. Peterková, K. Peltonen, N. A. Demopoulos, G. Stephanou, D. Vlachodimitropoulos, and R. J. Srám. 1996. "Biological effect monitoring in industrial workers from the Czech Republic exposed to low levels of butadiene." *Toxicology* 113 (1): 91-99. https://doi.org/https://doi.org/10.1016/0300-483X(96)03432-4. http://www.sciencedirect.com/science/article/pii/0300483X96034324.

Tice, R. R., E. Agurell, D. Anderson, B. Burlinson, A. Hartmann, H. Kobayashi, Y. Miyamae, E. Rojas, J. C. Ryu, and Y. F. Sasaki. 2000. "Single cell gel/comet assay: guidelines for *in vitro* and *in vivo* genetic toxicology testing." *Environ Mol Mutagen* 35 (3): 206-21.

Undeger, U., N. Basaran, A. Kars, and D. Guc. 1999. "Assessment of DNA damage in nurses handling antineoplastic drugs by the alkaline COMET assay." *Mutat Res* 439 (2): 277-85.

Undeger, U., A. F. Zorlu, and N. Basaran. 1999. "Use of the alkaline comet assay to monitor DNA damage in technicians exposed to low-dose radiation." *J Occup Environ Med* 41 (8): 693-8.

Uno, Y., H. Kojima, T. Omori, R. Corvi, M. Honma, L. M. Schechtman, R. R. Tice, C. Beevers, M. De Boeck, B. Burlinson, C. A. Hobbs, S. Kitamoto, A. R. Kraynak, J. McNamee, Y. Nakagawa, K. Pant, U. Plappert-Helbig, C. Priestley, H. Takasawa, K. Wada, U. Wirnitzer, N. Asano, P. A. Escobar, D. Lovell, T. Morita, M. Nakajima, Y. Ohno, and M. Hayashi. 2015. "JaCVAM-organized international validation study of the *in vivo* rodent alkaline comet assay for detection of genotoxic carcinogens: II. Summary of definitive validation study results." *Mutat Res Genet Toxicol Environ Mutagen* 786-788: 45-76. https://doi.org/10.1016/j.mrgentox.2015.04.010.

Valverde, M., and E. Rojas. 2009. "Environmental and occupational biomonitoring using the Comet assay." *Mutat Res* 681 (1): 93-109. https://doi.org/10.1016/j.mrrev.2008.11.001.

van Delft, J. H., M. S. Steenwinkel, J. G. van Asten, N. de Vogel, T. C. Bruijntjes-Rozier, T. Schouten, P. Cramers, L. Maas, M. H. van Herwijnen, F. van Schooten, and P. M. Hopmans. 2001. "Biological monitoring the exposure to polycyclic aromatic hydrocarbons of coke oven workers in relation to smoking and genetic polymorphisms for GSTM1 and GSTT1." *Ann Occup Hyg* 45 (5): 395-408.

Vijayalaxmi, R. R. Tice, and G. H. Strauss. 1992. "Assessment of radiation-induced DNA damage in human blood lymphocytes using the single-cell gel electrophoresis technique." *Mutat Res* 271 (3): 243-52.

Zhu, C. Q., T. H. Lam, and C. Q. Jiang. 2001. "Lymphocyte DNA damage in bus manufacturing workers." *Mutat Res* 491 (1-2): 173-81.

In: A Closer Look at the Comet Assay
Editor: Keith H. Harmon
ISBN: 978-1-53611-028-9
© 2019 Nova Science Publishers, Inc.

Chapter 3

THE COMET ASSAY AS A TOOL TO DETECT THE GENOTOXIC POTENTIAL OF NANOMATERIALS

Constanza Cortés[1,*] *and Ricard Marcos*[1,2,†]
[1]Group of Mutagenesis, Department of Genetics and Microbiology, Universitat Autònoma de Barcelona, Cerdanyola del Vallès, Spain
[2]CIBER Epidemiology and Public Health, ISCIII, Madrid, Spain

ABSTRACT

Since their discovery in the 1980s, the use of nanoparticles (NPs) has grown exponentially due to their distinctive physicochemical properties, which are exploited in fields as broad as electronics, medicine, food production, and packaging. However, this brings into question the potential toxicological issues derived from the increased exposure to NPs, especially DNA damage. It should be pointed out that, according to the relevance of the target, genotoxicity can determine undesired consequences such as immune responses or carcinogenicity. Out of the numerous assays utilized to detect NPs genotoxicity, the comet assay, an electrophoretic technique allowing the detection of DNA strand breaks in

---

[*] Corresponding Author's Email: constanza.cortes@uab.es.
[†] Corresponding Author's Email: ricard.marcos@uab.es.

single cells, is the most commonly used. This chapter focuses on its use to detect NPs-associated DNA damage, as well as its potential limitations due to the interaction of the NPs with the method. To achieve these goals, we present an overview of the available literature where this assay has been used to test NPs genotoxicity.

Keywords: comet assay, genotoxicity, nanomaterials, DNA damage, mammalian cells

INTRODUCTION

The Single-Cell Gel Electrophoresis Assay (SCGE), also known as the Alkaline Comet Assay, is one of the most used tests to evaluate the genotoxic damage in eukaryotic systems. It is a relatively simple technique, where isolated cells from previously-exposed plants, animals or cell cultures are mixed with agar, forming a gel that will be used to quantify the DNA damage of each cell. In its original version, cells embedded in agarose minigels over a microscope slide are lysed to remove membranes and histones, leaving a DNA-based structure called nucleoid. The gels are then subjected to electrophoresis, so DNA loops that present double-strand breaks (DSB) will extend more and form a "tail" towards the anode (Ostling & Johanson, 1984). The technique was soon improved by the introduction of strong alkaline conditions before and during the electrophoresis to induce DNA unwinding. This modification hugely increased the power of the assay by also allowing the detection of single-strand breaks (SSB) and alkali-labile sites (ALS) (Singh et al., 1988). Other modifications were subsequently added, like the incubation of gels with DNA repair enzymes such as formamidopyrimidine DNA glycosylase (FPG), which detects 8-oxoguanine and other purine oxidation adducts. FPG converts these lesions into DNA breaks, making the assay more multifaceted. In all its versions, the visualization by fluorescence microscopy of the "comets" and the intensity of their "tails" score the DNA damage suffered by each cell.

Since its development, there have been numerous publications backing up the usefulness of the comet assay to detect primary DNA damage and, recently, the standard version of the *in vivo* comet assay was published as an OECD guideline (OECD TG 489, 2014). The role of the *in vitro* version of the assay, however, has not been validated yet. Nonetheless, it has been accepted by the European Food Safety Authority (EFSA), and it is recommended as an appropriate test under the Registration, Evaluation, Authorization, and Restriction of Chemical Substances program of the European Commission (REACH) (Azqueta & Dusinska, 2015).

The use of the *in vitro* comet assay in assessing nanomaterials (NMs) safety has gained particular interest, as it avoids the ethical problems associated with animal research. These molecules present different physicochemical properties due to their distinct size (at least one dimension under 100 nm), shape, chemical composition, and surface charge and structure, therefore presenting different aggregation and solubility compared to their micro and macro counterparts (Collins et al., 2017). These differences could be crucial in their interaction with their cellular targets, eliciting an urgent necessity to analyze NMs genotoxic effects and toxicity mechanisms (Figure 1). However, NMs are very heterogeneous in nature, so the evaluation of the hazard they present to human beings cannot be generalized. Their interaction with biological systems does not only depend on the compound tested: different studies have shown that other parameters, such as their three-dimensional structure, chemical functionalization and the exposure time, define their interaction with the environment which, in turn, could determine their toxic effects. (García-Rodríguez et al., 2018; Saez-Tenorio et al., 2019; Tallec et al., 2019). Therefore, it is necessary to keep in mind that, when screening novel NMs to assess their genotoxic potential, one should analyze a vast number of conditions that mirror the different *in vivo* exposure scenarios. In the case of the traditional comet assay, this can be problematic, as multiple issues arise. On the one hand, the number of samples analyzed per experiment is limited to the electrophoresis tank, which can harbor 20 slides with two minigels each. On the other hand, the proper quantification of the damage normally needs the scoring of 50-100 comets per gel, which can request a high amount of time.

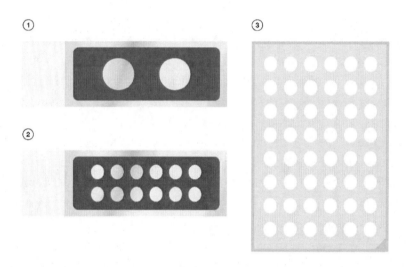

Figure 1. The original and modified comet assay supports. 1) Traditional comet assay slide, with two 70-100 µL agar drops. 2) Glass slide with twelve 4-7 µL agar drops. 3) 11 x 8 cm GelBond film, harboring forty-eight 4-7 µL agar drops.

To facilitate the task, a high throughput (HT) approach (the use of automated tools to rapidly analyze a large number of samples) could be of immense value. In the case of the comet assay, the most common modification is decreasing the size of the gels from 40 µL to 4 µL, allowing the fitting of 12 microgels on a glass slide, or 48 microgels on a GelBond film (Figure 2). Results demonstrated that the damage scored in the three different formats did not show statistical significance in any of the two different genotoxic agents tested (Azqueta et al., 2013). In addition, the use of automated or semi-automated scoring systems could diminish the analysis task. However, both systems present complications, as they depend on the accurate positioning of the gels on their support, an optimal embedding density with few comets overlapping, and low background fluorescence. Nonetheless, some companies have been able to develop automated scoring systems successfully. Comparative analyses of the different methods have shown that all three are capable of detecting significant damage levels, although visual scoring tends to overestimate low damage levels when compared with automated image analysis, while, heavily-damaged nucleoids are detected less efficiently with the automatic analysis (Azqueta et al., 2011).

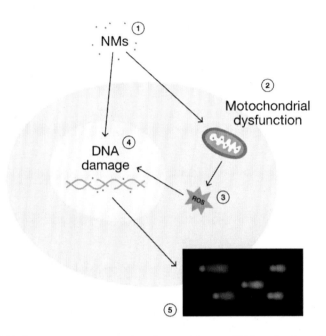

Figure 2. DNA damage assessment using the comet assay. NMs enter the cells by diffusion or active transport (1), where they can induce DNA damage by two mechanisms: disrupting mitochondrial metabolism (2) and/or producing ROS (3), or by directly interacting with DNA (4), producing single or double-strand breaks. Cells exposed to NMs are embedded in an agar matrix, where their cell membranes are lysed and the DNA is stripped from histones. The remaining nucleoids are subjected to electrophoresis, where the fragmented DNA strands migrate faster than unbroken strands to the anode, forming the tail of comet-like structures (5).

To date, the comet assay is the most used genotoxicity-assessing method for NMs. In this chapter, we aim to address the existing literature evaluating the genotoxic hazard of the most commonly-used NMs according to the Nanomaterial Consumer Products Inventory (CPI) (Vance et al., 2015). To this end, we have divided the studies according to the three most produced categories (metal, carbonaceous and silicon) and reviewed the literature considering only those studies performed on mammalian cells.

METAL AND METAL OXIDE NANOMATERIALS

Metal and metal oxide NMs are some of the most produced NMs, as they have applications that can be exploited for human consumables as well as industrial goods. Out of the 1814 products listed in the CPI, these NMs represent 37%. Among them, the most commonly-found metal-based NMs are silver (AgNPs), and the oxidized forms of titanium (titanium dioxide - TiO$_2$NPs) and zinc (zinc oxide - ZnONPs), respectively (Vance et al., 2015). Although silver is the most commonly used NM, TiO$_2$NPs holds the highest production rate. A 2012 report estimates the annual production of this nanomaterial to be 3000 tons/year, while AgNPs and ZnONPs production estimates are considerably lower, rounding the 55 and 550 tons/year, respectively (Piccinno et al., 2012). These high production volumes account for their use in a varied array of consumer products ranging from cosmetics to the food and supplements industry, mainly due to their antimicrobial and pigment properties, and, in the case of TiO$_2$NPs and ZnONPs, also their catalytic and semi-conductive activities, as well as UV filtering potential (Golbamaki et al., 2015). The main use of each NM determines the exposure route that would most-likely influence its entrance to the human body and, therefore, the genotoxic risk associated with each NM for each of the exposed tissues.

Genotoxicity of TiO$_2$NPs, as Assessed by the Comet Assay

TiO$_2$NPs are one of the most employed NM due to the multiple properties they acquire in the nano-size range. Among their vast applications in engineering, medicine, agriculture, food and cosmetic industry, their catalytic, pigment, and antibacterial properties are the most exploited. Given the vast use of this NM, the exposure routes range from inhalation, dermal absorption, ingestion and, to a lesser extent, injection, thus interacting with different organs and presenting a systemic hazard to humans. Previous studies have pointed out that exposure to TiO$_2$NPs can lead to digestion impairment, cardiovascular and brain tissue damage caused by oxidative stress, apoptosis and/or inflammatory response, which in turn can be associated to DNA damage and genetic instability (Baranowska-Wójcik et al., 2019). Much of the experimental genotoxicity data related to TiO$_2$NPs-induced DNA damage has been obtained from *in vitro* studies using the comet assay, as the technique can be easily performed in a wide array of cell types. Out of the sixty-one studies retrieved, only thirteen did not show a significant genotoxic effect under any of the assessed conditions, although oxidative DNA lesions were still observed in some of these studies when the FPG-modified version of the comet assay was used (Hackenberg et al., 2010, 2011, Vales et al., 2015; Schneider et al., 2017). Interestingly, the studies that gave negative genotoxic results were mainly performed in cells from the respiratory tract, which could suggest higher resistance of these epithelial cells to TiO$_2$NPs genotoxicity. Comparative studies between cells from different lineages also support this (Ursini et al., 2014; Cowie et al., 2015; El Yamani et al., 2017).

Another important aspect to consider when assessing TiO$_2$NPs genotoxicity is their three-dimensional structure and particle size. TiO$_2$ presents three crystalline structures: anatase, rutile and, less frequently, brookite. Out of the three, anatase has the strongest photocatalytic activity and has therefore been reckoned as the most toxic form (Weir et al., 2012). The enhanced toxicity of anatase crystals could be linked to their higher genotoxicity. Studies comparing different TiO$_2$NPs demonstrated that anatase DNA-damaging effects are stronger and usually linked to DNA oxidative damage (Gurr et al., 2005; Jugan et al., 2012; García-Rodríguez et al., 2018b; Louro et al., 2019). Regarding the effect of size over TiO$_2$NPs genotoxicity, two studies analyzing differences in genotoxicity between nano- and micro-sized TiO$_2$ demonstrate that particles in the nano-size range induce more pronounced genotoxic damage (Gurr et al., 2005; Hamzeh & Sunahara, 2013; Andreoli et al., 2018). However, comparative studies between different sizes under 100 nm are discordant: while some authors show a size-dependent reactivity (Falck et al., 2009; Jugan et al., 2012; Andreoli et al., 2018), others do not (Louro et al., 2019).

Finally, exposure conditions such as concentration and time should also be taken into account when assessing genotoxicity. Many of the studies taken into consideration show a direct relationship between genotoxic damage and TiO$_2$NPs concentration (Osman et al., 2010; Roszak et al., 2013; Schneider et al., 2017; Andreoli et al., 2018). However, the relationship between exposure time and genotoxicity is ambiguous. While some studies report a direct time-damage correlation (Jugan et al., 2012; Armand et al., 2016), others do not present such a clear relationship (Jugan et al., 2012; Joanna Roszak et al., 2013), and a number of studies have even reported DNA damage recovery after extended exposure periods (Jugan et al., 2012).

Taken together, the studies analyzing the genotoxic damage of TiO$_2$NPs using the comet assay indicate that this NM exerts genotoxic damage on most of the tested cell types, independently of the concentration and the exposure time analyzed.

Genotoxicity of ZnONPs, as Assessed by the Comet Assay

As the second most produced metal NM, ZnONPs also have a broad range of applications. Thanks to their ultraviolet protection, pigment, and antimicrobial properties, their main use is in personal care products like sunscreens and other topical products; paint, plastics, and dental care products. Furthermore, their catalytic activities make them suitable as biosensors and in electronic gadgets. The principal exposure route for this NM is inhalation, particularly in the case of occupational exposure of beauty products and paint production workers, followed by dermal contact and, due to their use as an antibacterial agent, ingestion. As with TiO$_2$NPs, ZnONPs can reach the bloodstream and be distributed to different organs, concentrating mainly on the lungs, liver, brain, kidney, bones, and spleen (Singh, 2019). This can lead to different outcomes such as inflammatory and immune responses, hepatic dysfunction, cardiopulmonary impairment and neuro- and nephrotoxicity, some of which are linked to DNA damage (Saptarshi et al., 2015). Once it has reached its target, ZnONPs can elicit cellular damage by three mechanisms: reactive oxygen species (ROS) production, zinc ions (Zn^{2+}) release, or direct interaction with different cell organelles. Considering the public health risk that this NM represents, the published data regarding ZnONPs *in vitro* toxicity is scarce and mainly focuses on assessing its cytotoxic impact. Among the consulted studies, only one study on HT29 cells reported that ZnONPs agglomerates in the nano-size range did not induce genotoxic damage (Schneider et al., 2017a). Otherwise, genotoxicity was observed in cells from different lineages, such as nasal epithelial cells (Hackenberg et al., 2017), lymphocytes (Naeem et al., 2018), adipose cells (Ickrath et al., 2017), epithelial intestinal cells (Caco-2) (Zijno et al., 2015), embryonic kidney cells (HEK293) (Demir, et al., 2014), lung (A54) and intestinal (HT29) epithelial cells, and human keratinocytes (Mu et al., 2014). Assessing the genotoxic effect over keratinocytes and epidermal cells, as well as possible synergic effects of ZnONPs and UV light, are of particular interest, given the widespread use of this NM in sunscreens lotions. Most of the studies using epidermal cells indicate a significant dose-dependent increase in genotoxicity, at least at exposures under 24 h, independently of the NPs size range (Sharma et al., 2009, 2011; Mu et al., 2014; Genç et al., 2018). As for studies focusing on UV light and ZnONPs interaction, a significant antagonistic effect was observed with UV-B light coexposure in HEK293 and NIH/3T3 fibroblasts (Demir et al., 2014).

Regarding the effect of exposure conditions over genotoxicity, conflicting results have been published. While most studies seem to agree that there is a concentration-dependent increase of genotoxic damage (Kononenko et al., 2017; Soni et al., 2017), there have been reports of direct and inverse correlation between exposure time and the genotoxicity elicited by ZnONPs (Bhattacharya et al., 2014; Zijno et al., 2015; El Yamani et al., 2017). A similar conflicting pattern exists regarding the importance of ZnONPs size in their genotoxic effect: even though it is generally accepted that a larger surface/mass ratio translates into higher reactivity, two of the four studies analyzing size relationship to genotoxicity gave negative results (Demir et al., 2014a, 2014b; Arakha et al., 2017; Naeem et al., 2018).

Summarizing, most studies assessing the genotoxic potential of ZnONPs agree that this nanomaterial elicits DNA damage, independently of the cell lineage. While an evident concentration-dependent increase in genotoxicity is observed in most reports, the effect of NP size and time exposure is not as explicit.

Genotoxicity of AgNPs, as Assessed by the Comet Assay

Ag is a valuable metal that has been used for millennia in the production of coins, jewelry, tableware and, in the past century, in photographic processing. However, in the last decades, its potent antimicrobial properties accounted to the production ROS and inactivation of bacterial enzymes have been exploited. The advance of the nanotechnology field has taken advantage of this feature, and AgNPs, especially those with sizes above 30 nm, can be found today in essential consumer goods such as food, personal care, textiles, and medical products (Flores-López et al., 2019). As such, the main route of exposure to this NP seems to be ingestion, followed by dermal contact, inhalation, and injection (Matteis, 2017). Although studies addressing the genotoxicity of AgNPs over the intestinal epithelial barrier did not show a genotoxic effect, a small but significant increase in oxidative DNA damage was observed using the FPG-modified version of the comet assay (Vila et al., 2018).

Exposure to Ag and AgNPs has been associated with different pathological effects such as anemia, growth retardation, cardiac enlargement, neurotoxicity, hepatotoxicity, and pulmonary inflammation liked to genotoxic and cytotoxic effects (Kim et al., 2013). This can be due to the high reactivity of Ag. Its high surface/volume ratio facilitates its ionization, increasing the production of ROS and making it more genotoxic than other NMs (Kermanizadeh et al., 2013). Its intrinsic instability also accounts for the high agglomeration and aggregation observed, which is considered a characteristic physicochemical property of these NPs. To address this issue, AgNPs are often coated with organic polymers to promote stability. Reports addressing the genotoxicity of pristine versus coated AgNPs indicate that the nature of the coating can either increase or reduce its genotoxicity (Ahamed et al., 2008; Wang et al., 2019). In fact, a comparative study between differently-coated AgNPs of similar sizes concluded that the polymer coat significantly changes the genotoxicity of the NPs (Brkić et al., 2017).

Different studies have discussed the importance of exposure conditions to AgNPs. Most of them agree that there is a time- and concentration-dependent increase in the genotoxic effect associated with this NPs (Wąsowicz et al., 2011; Juarez-Moreno et al., 2017; Wang et al., 2017). Also, although most of the published studies support the claim that smaller NP's sizes correlate to higher genotoxic damage due to an increased surface/mass ratio, there are reports giving contrasting results. While Schneider and Roszak observed an inverse correlation between size and the induced DNA damage, Lebedová and colleagues do not see any association (Schneider et al., 2017; Roszak et al., 2017; Lebedová et al., 2018).

Our overview of the published literature on AgNPs genotoxicity on mammalian cell culture models using the comet assay concludes that there are few publications addressing this issue. Nonetheless, most published findings corroborate the high DNA damage potential of AgNPs, either in their pristine or coated forms. Different parameters can increase the NPs' genotoxicity, such as the selected coating, longer exposure times, higher concentrations, and smaller NPs sizes. All this must be taken into account when designing AgNPs for consumer goods, considering the huge exposure that humans have to this NM.

SILICA NANOMATERIALS

Silica (SiO_2) is a naturally-occurring silicon dioxide that can exist in crystalline or colloidal form and comprises more than 10% of the planet's crust (Flörke et al., 2008). SiO_2NPs account for the highest production volume among manufactured NMs, with an estimate of 5500 tons/year that reflects its use in over 100 commercial products on the CPI. Due to its physicochemical properties, SiO$_2$NPs present a broad arrange of uses: they are widely used as excipients, viscosity-controlling, anti-foaming and anti-caking agents in the food industry, and as bulking, abrasive and suspending agents in cosmetics. Recently, they have gained attention in the biomedical and pharmaceutical fields due to their osteogenic properties as artificial implants components, and as drug delivery vehicles, thanks to their porous surface (Piccinno et al., 2012; Yazdimamaghani et al., 2019). Their main exposure routes for humans are dermal absorption, ingestion, inhalation, and to a lesser extent, injection. Estimates of SiO$_2$NPs oral intake are around 126 mg/day for a 70 kg person, which are well above the limits set by the Scientific Committee on Food of the European Food Safety Authority (20–50 mg/day for a 60 kg person) (Matteis, 2017). Such a high intake could be problematic, as the hazard posed by SiO$_2$ is well documented. On the one hand, exposure through inhalation to crystalline SiO$_2$ is toxic and can exert direct or indirect genotoxicity that leads to lung inflammation, systemic autoimmune diseases, and lung cancer, which prompted the International Agency for Research on Cancer (IARC) to classify crystalline silica as group 1 carcinogenic compound to humans. On the other hand, amorphous silica, the most used form of SiO$_2$NPs, is generally regarded as nontoxic by oral and topical applications and has not been recognized yet by IARC as a carcinogen (Kwon et al., 2014).

Even though the cytotoxicity of amorphous SiO$_2$NPs is well documented, the number of studies addressing the genotoxicity of these NPs using the comet assay is scarce. Albeit most authors observe DNA damage after cell cultures are exposed to silica NPs, others were unable to detect genotoxicity (Gehrke et al., 2012; Watson et al., 2014). However, the analyzed NPs presented low stability and larger sizes in the culture medium due to agglomeration, which could account for the lack of a genotoxic effect. Among the studies that did detect genotoxic damage, there seems to be a consensus on the importance of exposure conditions, as numerous studies demonstrate a dose-, and size-dependent effect. A study analyzing the genotoxicity of differently-sized SiO$_2$NPs, ranging from 10 to 100 nm in HUVEC cells, showed that independently of their diameter, there was a clear dose-dependent increment of the DNA damage observed. In addition, the olive tail moment (OTM) values of cells exposed to 10 and 25 nm particles were considerably higher than the OTM of cells exposed to 50 and 100 nm NPs (Zhou et al., 2019). Another study analyzing both the concentration- and size-dependent effects of the exposure to SiO$_2$NPs of 9, 15, 30 and 55nm on A549 cells obtained similar results (Maser et al., 2015).

Other important parameters that seem to influence silica NPs genotoxicity are the exposure time and the NPs functionalization. A 2017 study assessing the influence of functionalization over the genotoxic potential of different NMs showed that amino-SiO$_2$NPs and PEG-SiO$_2$NPs did not elicit genotoxic damage, but exposures to 50 µg/cm^2 unmodified and phosphate-SiO$_2$NPs significantly increased the DNA damage on EpiAirway™ 3D bronchial models (Haase et al., 2017). Furthermore, increasing the exposure time from 24 to 60 h increased the genotoxic potential of unmodified SiO$_2$NPs, suggesting that increasing time exposures also has a detrimental effect on DNA integrity. This study stresses the need for more studies investigating the genotoxicity of different functionalization, considering the potential use of silica NPs as a drug carrier. Interestingly, only one study addressed the genotoxic effect of the interaction of nano-silica with a known genotoxic molecule (Lu et al., 2015). While the exposure of A549 cells to 10 µg/mL of 40 nm SiO$_2$NPs had no effect over DNA damage, the coexposure with 100 µM of lead acetate, a known genotoxic agent, significantly doubled the genotoxicity of lead exposure alone.

Overall, even if the number of studies describing the genotoxic consequences of SiO$_2$NPs is scarce, several factors appear to modulate its DNA-damaging effects, such as concentration, exposure time, NP size and functionalization of the core molecule. However, more studies are required to better characterize the impact of these factors over the genotoxic effects of nano-silica alone or as a genotoxicity-potentiating agent over DNA-damaging compounds.

CARBONACEOUS NANOMATERIALS

Carbonaceous NPs are found in the environment as a component of air pollution and as a fraction of naturally-occurring phenomena, such as volcanic eruptions. However, engineered carbon-based nanomaterials (CNM) have gained popularity over the past decade due to their unique features. Carbon's atomic structure grants physical, chemical, and hybridization properties that allow a much more customizable manipulation when compared to inorganic NMs. This makes them excellent candidates for biomedical applications such as diagnosis, imaging, targeted photothermal therapy, drug delivery, and tissue engineering, as well as multifaceted materials for industrial use. Their shape, size, and surface chemistry are important parameters that determine their electronic, optical, thermal, and sorption properties, designating the potential use of each CNM (Mauter et al., 2008; Chaudhuri et al., 2018). Nonetheless, this enhanced malleability could account for a higher reactivity in biological systems. Considering that most of the potential uses are related to medical applications, the characterization of potential pathologic effects of CNM, such as DNA damage, is essential to further explore the possibilities these materials could open in the biomedical field.

Genotoxicity of Carbon Nanotubes as Assessed by the Comet Assay

Carbon nanotubes (CNT) are one of the most produced CNM, with a median production rate of 300 tons per year and, as such, one of the most studied (Piccinno et al., 2012). They are one-dimensional materials formed by one or more graphite monoatomic sheets rolled into a cylinder of nanoscale diameter and micrometer length, which is usually capped with a spherical fullerene. Depending on the number of graphite sheets that form the CNT's walls, they can be classified as single-walled carbon nanotubes (SWCNT) or multi-walled carbon nanotubes (MWCNT) (Zhang et al., 2014). Their flexibility allows them to easily cross the cell membrane either by a spiraling motion or by interacting with proteins, which makes them promising agents for biomedical research, biosensing, imaging, drug delivery, photothermal therapy, and tissue engineering purposes. However, assessing the safety of CNT use is difficult to address, as CNT's toxicity and behavior are related to different properties such as their structure, number of walls, diameter, and length, among others (Simon et al., 2019). Human exposure to CNT may happen through different pathways: oral, injection, inhalation and dermal contact. Although few studies have focused on the toxicity related to the entry route of CNT, it has been reported that injection, oral, and dermal administration cause mild inflammation, but their inhalation may result in severe inflammation that could induce asthma, bronchitis, emphysema, and lung cancer (Madani et al., 2013). At a cellular level, *in vitro* studies using the comet assay to assess DNA damage show increased levels of strand breaks, although the opposite effect has also been reported. For example, a 2015 report analyzing the difference in genotoxicity between two differently-shaped MWCNT in BEAS-2B cells demonstrated that straight nanotubes are more genotoxic than their tangled counterparts (Catalán et al., 2016). Another 2013 study analyzing the genotoxicity of two MWCNT of different lengths observed a significant dose-response dependent genotoxic effect on HK-2 cells, which was associated with DNA oxidative damage (Kermanizadeh et al., 2013). However, a similar study using the same NMs and exposure conditions did not elicit DNA damage in three different cell lines, suggesting that differences in the preparation and experimental protocols could have a great impact on the outcome of the assessment (Thongkam et al., 2017). Studies focusing on the genotoxic effects of prolonged exposure times of pristine MWCNT on BEAS-2B cells also failed to detect genotoxic and DNA oxidative response, although ROS production and other cellular markers associated with transformation were observed (Vales et al., 2016). Similarly, a genotoxic analysis of fifteen pristine or differently-functionalized CNT of different diameters on murine epithelial lung cells only induced DNA damage on -COOH functionalized MWCNT, and only at the highest concentration tested (Jackson et al., 2015). The difference of the results observed in these studies does not allow for a concluding verdict on CNT's genotoxicity, stressing the need for further investigation and homogenization of the protocols used to assess CNT genotoxic effects.

Genotoxicity of Graphene, as Assessed by the Comet Assay

Graphene is a two-dimensional CNM formed by hexagonally-arranged carbon atoms. The basic structure of graphene can be modified to form a wide variety of materials, which include its pristine, one-layer form (SLG), few-layer graphene (FLG), multilayer graphene (MLG), graphene oxide (GO), reduced graphene oxide (rGO), graphene nanosheets (GNS), and graphene nanoplatelets (GNP). Graphene-based materials present a larger surface/mass ratio and better dispersibility than CNT, which grants this family of CNM an enormous potential for biomedical applications. However, graphene is newer and less characterized than its tubular counterpart, making the evaluation of their potential cytotoxic and genotoxic effects an open and virtually unexplored field of research. Furthermore, the unique physicochemical properties that each of the different forms of graphene present could generate different toxicological effects, inciting the urgent necessity to establish a strategy for its toxicity evaluation (Seabra et al., 2014).

To date, very few studies have analyzed the genotoxic effects of graphene-based NMs using the comet assay. However, most of them indicate that these NMs have DNA damage potential. Akhavan and colleagues evaluated the influence of size and concentration of rGO of on human mesenchymal stem cells (hMSCs) showing a clear dose- and size-dependent trend, as the smaller nanoparticles were significantly more genotoxic than then larger counterparts (Akhavan et al., 2012). The oxidation state and functionalization of graphene also seems to play a role in determining its genotoxicity. A 2016 study compared the DNA damage potential of pristine and functionalized GNPs, as well as single-layer and few-layer GO on BEAS-2B cells (Chatterjee et al., 2016). Results showed that all the different forms of graphene exerted genotoxicity in a dose-dependent manner and that GNPs were more genotoxic than GO. Furthermore, the different functionalization groups and a different number of layers also determined different levels of DNA damage. The genotoxic potential was higher for pristine GNPs, followed by COOH-GNPs, NH_2-GNPs, FLGO, and SLGO. The genotoxic potential of GO has been proposed as a useful tool for cancer therapy. Krasteva and colleagues treated the mouse colorectal cancer cell line Colon 26 with mono- and bilayered pristine and amino-functionalized GO to assess their antitumoral potential (Krasteva et al., 2019). While results showed a strong cytotoxic effect for both forms of GO after exposures lasting for 24 h, the comet assay showed that only NH_2-GO exerted a genotoxic effect. The authors account the inability to detect genotoxic damage on the high apoptotic rate of the samples, which was nearly 70% and 100% for NH-GO and GO, respectively.

In summary, the current literature addressing graphene, GO and rGO effects on cell cultures prove their potential to trigger cytotoxic and genotoxic effects. However, more studies are needed to determine the hazard that these nanomaterials pose and if they could be safely used as biomedical products.

CONCLUSION

The applications of organic and inorganic NMs have developed greatly in the past few decades, leading to their use in a vast number of commodities such as food, cosmetics, biomedical and other products. For this reason, the need to have a comprehensive understanding of their adverse effects when interacting with living systems is essential for their safe use, especially the detrimental effects that NMs could have over DNA's stability. Most of the literature evaluating the genotoxic effects of the widely-used and up-and-coming NMs using the comet assay indicates that they exert genotoxic damage in cell culture systems. However, there seem to be some contrasting results regarding this issue, as many studies also fail to observe genotoxic damage. The lack of a standardized protocol for the assessment of genotoxicity with the comet assay on cell cultures limits the possibility to compare between studies, as many present different synthesis and characterization methodologies, culture and exposure conditions and statistical analyses. The application of a standardized protocol would certainly help to ascertain the real DNA damage potential of NMs. Also, the implementation of systematic studies evaluating the genotoxicity of multiple NMs with different physicochemical properties, such as the high-throughput comet assay, would allow a more reliable comparison of their genotoxic features. Also, some authors have raised the question that the increased reactivity of NMs, when compared to their bulk form, could create artifacts in the assay, as their direct interaction with DNA or with the FPG enzyme, or their capacity to promote ROS formation due to their photocatalytic activity could create artifacts that give false-positives or false-negatives (Karlsson et al., 2015). All these obstacles should be addressed to improve the reliability of the results obtained to fully characterize the real hazard that NMs pose to human health.

REFERENCES

Ahamed, M., Karns, M., Goodson, M., Rowe, J., Hussain, S. M., Schlager, J. J., & Hong, Y. (2008). DNA damage response to different surface chemistry of silver nanoparticles in mammalian cells. *Toxicology and Applied Pharmacology*, *233*(3), 404–410. https://doi.org/10.1016/j.taap.2008.09.015.

Akhavan, O., Ghaderi, E., & Akhavan, A. (2012). Size-dependent genotoxicity of graphene nanoplatelets in human stem cells. *Biomaterials*, *33*(32), 8017–8025. https://doi.org/10.1016/j.biomaterials.2012.07.040.

Andreoli, C., Leter, G., De Berardis, B., Degan, P., De Angelis, I., Pacchierotti, F., Crebelli, R., Barone, F., & Zijno, A. (2018). Critical issues in genotoxicity assessment of TiO_2 nanoparticles by human peripheral blood mononuclear cells. *Journal of Applied Toxicology*, *38*(12), 1471–1482. https://doi.org/10.1002/jat.3650.

Arakha, M., Roy, J., Nayak, P. S., Mallick, B., & Jha, S. (2017). Zinc oxide nanoparticle energy band gap reduction triggers the oxidative stress resulting into autophagy-mediated apoptotic cell death. *Free Radical Biology and Medicine*, *110*, 42–53. https://doi.org/10.1016/j.freeradbiomed.2017.05.015.

Armand, L., Tarantini, A., Beal, D., Biola-Clier, M., Bobyk, L., Sorieul, S., Pernet-Gallay, K., Marie-Desvergne, C., Lynch, I., Herlin-Boime, N., & Carriere, M. (2016). Long-term exposure of A549 cells to titanium dioxide nanoparticles induces DNA damage and sensitizes cells towards genotoxic agents. *Nanotoxicology*, *10*(7), 913–923. https://doi.org/10.3109/17435390.2016.1141338.

Azqueta, A., Meier, S., Priestley, C., Gutzkow, K. B., Brunborg, G., Sallette, J., Soussaline, F., & Collins, A. (2011). The influence of scoring method on variability in results obtained with the comet assay. *Mutagenesis*, *26*(3), 393–399. https://doi.org/10.1093/mutage/geq105.

Azqueta, Amaya, & Dusinska, M. (2015). The use of the comet assay for the evaluation of the genotoxicity of nanomaterials. *Frontiers in Genetics*, *6*, 239. https://doi.org/10.3389/fgene.2015.00239.

Azqueta, Amaya, Gutzkow, K. B., Priestley, C. C., Meier, S., Walker, J. S., Brunborg, G., & Collins, A. R. (2013). A comparative performance test of standard, medium- and high-throughput comet assays. *Toxicology in Vitro*, *27*(2), 768–773. https://doi.org/10.1016/j.tiv.2012.12.006.

Baranowska-Wójcik, E., Szwajgier, D., Oleszczuk, P., & Winiarska-Mieczan, A. (2019). Effects of Titanium Dioxide Nanoparticles Exposure on Human Health—a Review. *Biological Trace Element Research*. [Epub ahead of print]. https://doi.org/10.1007/s12011-019-01706-6.

Bhattacharya, D., Santra, C. R., Ghosh, A. N., & Karmakar, P. (2014). Differential Toxicity of Rod and Spherical Zinc Oxide Nanoparticles on Human Peripheral Blood Mononuclear Cells. *Journal of Biomedical Nanotechnology*, *10*(4), 707–716. https://doi.org/10.1166/jbn.2014.1744.

Brkić Ahmed, L., Milić, M., Pongrac, I. M., Marjanović, A. M., Mlinarić, H., Pavičić, I., Gajović, S., Vinković Vrček, I. (2017). Impact of surface functionalization on the uptake mechanism and toxicity effects of silver nanoparticles in HepG2 cells. *Food and Chemical Toxicology*, *107*(Pt A), 349–361. https://doi.org/10.1016/j.fct.2017.07.016.

Catalán, J., Siivola, K. M., Nymark, P., Lindberg, H., Suhonen, S., Järventaus, H., Koivisto, A.J., Moreno, C., Vanhala, E., Wolff, H., Kling, K. I., Jensen, K. A., Savolainen, K., & Norppa, H. (2016). *In vitro* and *in vivo* genotoxic effects of straight versus tangled multi-walled carbon nanotubes. *Nanotoxicology*, *10*(6), 794–806. https://doi.org/10.3109/17435390.2015.1132345.

Chatterjee, N., Yang, J., & Choi, J. (2016). Differential genotoxic and epigenotoxic effects of graphene family nanomaterials (GFNs) in human bronchial epithelial cells. *Mutation Research/Genetic Toxicology and Environmental Mutagenesis*, *798–799*, 1–10. https://doi.org/10.1016/j.mrgentox.2016.01.006.

Chaudhuri, I., Fruijtier-Pölloth, C., Ngiewih, Y., & Levy, L. (2018). Evaluating the evidence on genotoxicity and reproductive toxicity of carbon black: a critical review. *Critical Reviews in Toxicology*, *48*(2), 143–169. https://doi.org/10.1080/10408444.2017.1391746.

Chen, Z., Wang, Y., Ba, T., Li, Y., Pu, J., Chen, T., Song, Y., Gu, Y., Qian, Q., Yang, J., & Jia, G. (2014). Genotoxic evaluation of titanium dioxide nanoparticles in vivo and in vitro. *Toxicology Letters*, *226*(3), 314–319. https://doi.org/10.1016/j.toxlet.2014.02.020.

Collins, A. R., Annangi, B., Rubio, L., Marcos, R., Dorn, M., Merker, C., Estrela-Lopis, I., Cimpan, M.R., Ibrahim, M., Cimpan, E., Ostermann, M., Sauter, A., Yamani, N. E., Shaposhnikov, S., Chevillard, S., Paget, V., Grall, R., Delic, J., de-Cerio F. G., Suarez-Merino, B., Fessard, V., Hogeveen, K. N., Fjellsbø, L. M., Pran, E. R., Brzicova, T., Topinka, J., Silva, M. J., Leite, P. E., Ribeiro, A. R., Granjeiro, J. M., Grafström, R., Prina-Mello, A., & Dusinska, M. (2017). High throughput toxicity screening and intracellular detection of nanomaterials. *Wiley Interdisciplinary Reviews. Nanomedicine and Nanobiotechnology*, *9*(1). https://doi.org/10.1002/wnan.1413.

Cowie, H., Magdolenova, Z., Saunders, M., Drlickova, M., Correia Carreira, S., Halamoda Kenzaoi, B., Gombau, L., Guadagnini, R., Lorenzo, Y., Walker, L., Fjellsbø, L. M., Huk, A., Rinna, A., Tran, L., Volkovova, K., Boland, S., Juillerat-Jeanneret, L., Marano, F., Collins, A.R., & Dusinska, M. (2015). Suitability of human and mammalian cells of different origin for the assessment of genotoxicity of metal and polymeric engineered nanoparticles. *Nanotoxicology*, *9 Suppl 1*, 57–65. https://doi.org/10.3109/17435390.2014.940407.

De Matteis, V. (2017). Exposure to Inorganic Nanoparticles: Routes of Entry, Immune Response, Biodistribution and In Vitro/In Vivo Toxicity Evaluation. *Toxics*, *5*(4). https://doi.org/10.3390/TOXICS 5040029.

Demir, E., Akça, H., Kaya, B., Burgucu, D., Tokgün, O., Turna, F., Aksakal, S., Vales, G., Creus, A., & Marcos, R. (2014a). Zinc oxide nanoparticles: Genotoxicity, interactions with UV-light and cell-transforming potential. *Journal of Hazardous Materials*, *264*, 420–429. https://doi.org/10.1016/j.jhazmat.2013.11.043.

Demir, E., Creus, A., & Marcos, R. (2014b). Genotoxicity and DNA Repair Processes of Zinc Oxide Nanoparticles. *Journal of Toxicology and Environmental Health, Part A*, *77*(21), 1292–1303. https://doi.org/10.1080/15287394.2014.935540.

El Yamani, N., Collins, A. R., Rundén-Pran, E., Fjellsbø, L. M., Shaposhnikov, S., Zielonddiny, S., & Dusinska, M. (2017). In vitro genotoxicity testing of four reference metal nanomaterials, titanium dioxide, zinc oxide, cerium oxide and silver: towards reliable hazard assessment. *Mutagenesis*, *32*(1), 117–126. https://doi.org/10.1093/mutage/gew060.

Falck, G. C. M., Lindberg, H. K., Suhonen, S., Vippola, M., Vanhala, E., Catalán, J., Savolainen, K., & Norppa, H. (2009). Genotoxic effects of nanosized and fine TiO_2. *Human & Experimental Toxicology*, *28*(6–7), 339–352. https://doi.org/10.1177/0960327109105163.

Flores-López, L. Z., Espinoza-Gómez, H., & Somanathan, R. (2019). Silver nanoparticles: Electron transfer, reactive oxygen species, oxidative stress, beneficial and toxicological effects. Mini review.

Journal of Applied Toxicology, *39*(1), 16–26. https://doi.org/10.1002/jat.3654.

Flörke, O. W., Graetsch, H. A., Brunk, F., Benda, L., Paschen, S., Bergna, H. E., & Schiffmann, D. (2008). Silica. *Ullmann's Encyclopedia of Industrial Chemistry*. Weinheim, Germany: Wiley-VCH, Verlag GmbH & Co. KGaA. https://doi.org/10.1002/14356007.a23_583.pub3.

García-Rodríguez, A., Vila, L., Cortés, C., Hernández, A., & Marcos, R. (2018). Effects of differently shaped TiO$_2$NPs (nanospheres, nanorods and nanowires) on the in vitro model (Caco-2/HT29) of the intestinal barrier. *Particle and Fibre Toxicology*, *15*(1), 33. https://doi.org/10.1186/s12989-018-0269-x.

Gehrke, H., Frühmesser, A., Pelka, J., Esselen, M., Hecht, L. L., Blank, H., Schuchmann, H. P., Gerthsen, D., Marquardt, C., Diabaté, S., Weiss, C., & Marko, D. (2012). In vitro toxicity of amorphous silica nanoparticles in human colon carcinoma cells. *Nanotoxicology*, *7*(3), 274–293. https://doi.org/10.3109/17435390.2011.652207.

Genç, H., Barutca, B., Koparal, A. T., Özöğüt, U., Şahin, Y., & Suvacı, E. (2018). Biocompatibility of designed MicNo-ZnO particles: Cytotoxicity, genotoxicity and phototoxicity in human skin keratinocyte cells. *Toxicology in Vitro : An International Journal Published in Association with BIBRA*, *47*, 238–248. https://doi.org/10.1016/j.tiv.2017.12.004.

Golbamaki, N., Rasulev, B., Cassano, A., Marchese Robinson, R. L., Benfenati, E., Leszczynski, J., & Cronin, M. T. D. (2015). Genotoxicity of metal oxide nanomaterials: review of recent data and discussion of possible mechanisms. *Nanoscale*, *7*(6), 2154–2198. https://doi.org/10.1039/c4nr06670g.

Gurr, J. R., Wang, A. S. S., Chen, C. H., & Jan, K. Y. (2005). Ultrafine titanium dioxide particles in the absence of photoactivation can induce oxidative damage to human bronchial epithelial cells. *Toxicology*, *213*(1–2), 66–73. https://doi.org/10.1016/j.tox.2005.05.007.

Haase, A., Dommershausen, N., Schulz, M., Landsiedel, R., Reichardt, P., Krause, B. C., Tentschert, J., & Luch, A. (2017). Genotoxicity testing of different surface-functionalized SiO$_2$, ZrO$_2$ and silver nanomaterials in 3D human bronchial models. *Archives of Toxicology*, *91*(12), 3991–4007. https://doi.org/10.1007/s00204-017-2015-9.

Hackenberg, S., Friehs, G., Froelich, K., Ginzkey, C., Koehler, C., Scherzed, A., Burghartz, M., Hagen, R., & Kleinsasser, N. (2010). Intracellular distribution, geno- and cytotoxic effects of nanosized titanium dioxide particles in the anatase crystal phase on human nasal mucosa cells. *Toxicology Letters*, *195*(1), 9–14. https://doi.org/10.1016/j.toxlet.2010.02.022.

Hackenberg, S., Friehs, G., Kessler, M., Froelich, K., Ginzkey, C., Koehler, C., Scherzed, A., Burghartz, M., & Kleinsasser, N. (2011). Nanosized titanium dioxide particles do not induce DNA damage in human peripheral blood lymphocytes. *Environmental and Molecular Mutagenesis*, *52*(4), 264–268. https://doi.org/10.1002/em.20615.

Hackenberg, S., Scherzed, A., Zapp, A., Radeloff, K., Ginzkey, C., Gehrke, T., Ickrath, P., & Kleinsasser, N. (2017). Genotoxic effects of zinc oxide nanoparticles in nasal mucosa cells are antagonized by titanium dioxide nanoparticles. *Mutation Research. Genetic Toxicology and Environmental Mutagenesis*, *816–817*, 32–37. https://doi.org/10.1016/j.mrgentox.2017.02.005.

Hamzeh, M., & Sunahara, G. I. (2013). In vitro cytotoxicity and genotoxicity studies of titanium dioxide (TiO$_2$) nanoparticles in Chinese hamster lung fibroblast cells. *Toxicology in Vitro*, *27*(2), 864–873. https://doi.org/10.1016/j.tiv.2012.12.018.

Ickrath, P., Wagner, M., Scherzad, A., Gehrke, T., Burghartz, M., Hagen, R., Radeloff, K., Kleinsasser, N., & Hackenberg, S. (2017). Time-Dependent Toxic and Genotoxic Effects of Zinc Oxide Nanoparticles after Long-Term and Repetitive Exposure to Human Mesenchymal Stem Cells. *International Journal of Environmental Research and Public Health*, *14*(12), 1590. https://doi.org/10.3390/ijerph14121590.

Jackson, P., Kling, K., Jensen, K. A., Clausen, P. A., Madsen, A. M., Wallin, H., & Vogel, U. (2015). Characterization of genotoxic response to 15 multiwalled carbon nanotubes with variable physico-chemical properties including surface functionalizations in the FE1-

Muta(TM) mouse lung epithelial cell line. *Environmental and Molecular Mutagenesis*, *56*(2), 183–203. https://doi.org/10.1002/em.21922.

Juarez-Moreno, K., Gonzalez, E., Girón-Vazquez, N., Chávez-Santoscoy, R., Mota-Morales, J., Perez-Mozqueda, L. L., Garcia-Garcia, M. R., Pestryakov, A., & Bogdanchikova, N. (2017). Comparison of cytotoxicity and genotoxicity effects of silver nanoparticles on human cervix and breast cancer cell lines. *Human & Experimental Toxicology*, *36*(9), 931–948. https://doi.org/10.1177/0960327116675206.

Jugan, M. L., Barillet, S., Simon-Deckers, A., Herlin-Boime, N., Sauvaigo, S., Douki, T., & Carriere, M. (2012). Titanium dioxide nanoparticles exhibit genotoxicity and impair DNA repair activity in A549 cells. *Nanotoxicology*, *6*(5), 501–513. https://doi.org/10.3109/17435390.2011.587903.

Karlsson, H. L., Di Bucchianico, S., Collins, A. R., & Dusinska, M. (2015). Can the comet assay be used reliably to detect nanoparticle-induced genotoxicity? *Environmental and Molecular Mutagenesis*, *56*(2), 82–96. https://doi.org/10.1002/em.21933.

Kermanizadeh, A., Vranic, S., Boland, S., Moreau, K., Baeza-Squiban, A., Gaiser, B. K., Andrzejczuk, L. A., & Stone, V. (2013). An in vitroassessment of panel of engineered nanomaterials using a human renal cell line: cytotoxicity, pro-inflammatory response, oxidative stress and genotoxicity. *BMC Nephrology*, *14*(1), 96. https://doi.org/10.1186/1471-2369-14-96.

Kim, S., & Ryu, D. Y. (2013). Silver nanoparticle-induced oxidative stress, genotoxicity and apoptosis in cultured cells and animal tissues. *Journal of Applied Toxicology*, *33*(2), 78–89. https://doi.org/10.1002/jat.2792.

Kononenko, V., Repar, N., Marušič, N., Drašler, B., Romih, T., Hočevar, S., & Drobne, D. (2017). Comparative in vitro genotoxicity study of ZnO nanoparticles, ZnO macroparticles and $ZnCl_2$ to MDCK kidney cells: Size matters. *Toxicology in Vitro*, *40*, 256–263. https://doi.org/10.1016/j.tiv.2017.01.015.

Krasteva, N., Keremidarska-Markova, M., Hristova-Panusheva, K., Andreeva, T., Speranza, G., Wang, D., Draganova-Filipova, M., Miloshev, G., & Georgieva, M. (2019). Aminated Graphene Oxide as a

Potential New Therapy for Colorectal Cancer. *Oxidative Medicine and Cellular Longevity, 2019*, 1–15. https://doi.org/10.1155/2019/3738980.

Kwon, J. Y., Koedrith, P., & Seo, Y. R. (2014). Current investigations into the genotoxicity of zinc oxide and silica nanoparticles in mammalian models in vitro and in vivo: carcinogenic/genotoxic potential, relevant mechanisms and biomarkers, artifacts, and limitations. *International Journal of Nanomedicine, 9 Suppl 2*, 271. https://doi.org/10.2147/IJN.S57918.

Lebedová, J., Hedberg, Y. S., Odnevall Wallinder, I., & Karlsson, H. L. (2018). Size-dependent genotoxicity of silver, gold and platinum nanoparticles studied using the mini-gel comet assay and micronucleus scoring with flow cytometry. *Mutagenesis, 33*(1), 77–85. https://doi.org/10.1093/mutage/gex027.

Louro, H., Saruga, A., Santos, J., Pinhão, M., & Silva, M. J. (2019). Biological impact of metal nanomaterials in relation to their physicochemical characteristics. *Toxicology in Vitro, 56*, 172–183. https://doi.org/10.1016/j.tiv.2019.01.018.

Lu, C. F., Yuan, X. Y., Li, L. Z., Zhou, W., Zhao, J., Wang, Y. M., & Peng, S. Q. (2015). Combined exposure to nano-silica and lead induced potentiation of oxidative stress and DNA damage in human lung epithelial cells. *Ecotoxicology and Environmental Safety, 122*, 537–544. https://doi.org/10.1016/j.ecoenv.2015.09.030.

Madani, S. Y., Mandel, A., & Seifalian, A. M. (2013). A concise review of carbon nanotube's toxicology. *Nano Reviews, 4*. https://doi.org/10.3402/nano.v4i0.21521.

Maser, E., Schulz, M., Sauer, U. G., Wiemann, M., Ma-Hock, L., Wohlleben, W., Hartwig, A., Landsiedel, R. (2015). In vitro and in vivo genotoxicity investigations of differently sized amorphous SiO_2 nanomaterials. *Mutation Research/Genetic Toxicology and Environmental Mutagenesis, 794*, 57–74. https://doi.org/10.1016/j.mrgentox.2015.10.005.

Mauter, M. S., & Elimelech, M. (2008). Environmental applications of carbon-based nanomaterials. *Environmental Science & Technology*,

42(16), 5843–5859. Retrieved from http://www.ncbi.nlm.nih.gov/pubmed/18767635.

Mu, Q., David, C. A., Galceran, J., Rey-Castro, C., Krzemiński, Ł., Wallace, R., Bamiduro, F., Milne, S. J., Hondow, N. S., Brydson, R., Vizcay-Barrena, G., Routledge, M. N., Jeuken, L. J., & Brown, A. P. (2014). Systematic Investigation of the Physicochemical Factors That Contribute to the Toxicity of ZnO Nanoparticles. *Chemical Research in Toxicology*, *27*(4), 558–567. https://doi.org/10.1021/tx4004243.

Naeem, A., Alam, M. T., Khan, T. A., & Husain, Q. (2018). A Biophysical and Computational Study of Concanavalin A Immobilized Zinc Oxide Nanoparticles. *Protein and Peptide Letters*, *24*(12), 1096–1104. https://doi.org/10.2174/0929866524666170920114057.

OECD TG 489. (2014). *Test No. 489: In Vivo Mammalian Alkaline Comet Assay. OECD. In: OECD Guidelines for testing of chemicals.* OECD. https://doi.org/10.1787/9789264224179-en.

Ostling, O., & Johanson, K. J. (1984). Microelectrophoretic study of radiation-induced DNA damages in individual mammalian cells. *Biochemical and Biophysical Research Communications*, *123*(1), 291–298. Retrieved from http://www.ncbi.nlm.nih.gov/pubmed/6477583.

Piccinno, F., Gottschalk, F., Seeger, S., & Nowack, B. (2012). Industrial production quantities and uses of ten engineered nanomaterials in Europe and the world. *Journal of Nanoparticle Research*, *14*(9), 1109. https://doi.org/10.1007/s11051-012-1109-9.

Roszak, J., Domeradzka-Gajda, K., Smok-Pieniążek, A., Kozajda, A., Spryszyńska, S., Grobelny, J., Tomaszewska, E., Ranoszek-Soliwoda, K., Cieślak, M., Puchowicz D., & Stępnik, M. (2017). Genotoxic effects in transformed and non-transformed human breast cell lines after exposure to silver nanoparticles in combination with aluminium chloride, butylparaben or di- n -butylphthalate. *Toxicology in Vitro*, *45*(Pt 1), 181–193. https://doi.org/10.1016/j.tiv.2017.09.003.

Roszak, Joanna, Stępnik, M., Nocuń, M., Ferlińska, M., Smok-Pieniążek, A., Grobelny, J., Tomaszewska, E., Wąsowicz, W., & Cieślak, M. (2013). A strategy for in vitro safety testing of nanotitania-modified textile products. *Journal of Hazardous Materials*, *256–257*, 67–75. https://doi.org/10.1016/j.jhazmat.2013.04.022.

Saez-Tenorio, M., Domenech, J., García-Rodríguez, A., Velázquez, A., Hernández, A., Marcos, R., & Cortés, C. (2019). Assessing the relevance of exposure time in differentiated Caco-2/HT29 cocultures. Effects of silver nanoparticles. *Food and Chemical Toxicology*, *123*, 258–267. https://doi.org/10.1016/j.fct.2018.11.009.

Saptarshi, S. R., Duschl, A., & Lopata, A. L. (2015). Biological reactivity of zinc oxide nanoparticles with mammalian test systems: an overview. *Nanomedicine*, *10*(13), 2075–2092. https://doi.org/10.2217/nnm.15.44.

Schneider, T., Westermann, M., & Glei, M. (2017b). In vitro uptake and toxicity studies of metal nanoparticles and metal oxide nanoparticles in human HT29 cells. *Archives of Toxicology*, *91*(11), 3517–3527. https://doi.org/10.1007/s00204-017-1976-z.

Seabra, A. B., Paula, A. J., de Lima, R., Alves, O. L., & Durán, N. (2014). Nanotoxicity of Graphene and Graphene Oxide. *Chemical Research in Toxicology*, *27*(2), 159–168. https://doi.org/10.1021/tx400385x.

Sharma, V., Anderson, D., & Dhawan, A. (2011). Zinc oxide nanoparticles induce oxidative stress and genotoxicity in human liver cells (HepG2). *Journal of Biomedical Nanotechnology*, *7*(1), 98–99. Retrieved from http://www.ncbi.nlm.nih.gov/pubmed/21485822.

Sharma, V., Shukla, R. K., Saxena, N., Parmar, D., Das, M., & Dhawan, A. (2009). DNA damaging potential of zinc oxide nanoparticles in human epidermal cells. *Toxicology Letters*, *185*(3), 211–218. https://doi.org/10.1016/J.TOXLET.2009.01.008.

Simon, J., Flahaut, E., & Golzio, M. (2019). Overview of Carbon Nanotubes for Biomedical Applications. *Materials*, *12*(4), 624. https://doi.org/10.3390/ma12040624.

Singh, N. P., McCoy, M. T., Tice, R. R., & Schneider, E. L. (1988). A simple technique for quantitation of low levels of DNA damage in individual cells. *Experimental Cell Research*, *175*(1), 184–191.

Singh, S. (2019). Zinc oxide nanoparticles impacts: cytotoxicity, genotoxicity, developmental toxicity, and neurotoxicity. *Toxicology*

Mechanisms and Methods, *29*(4), 300–311. https://doi.org/10.1080/15376516.2018.1553221.

Soni, D., Gandhi, D., Tarale, P., Bafana, A., Pandey, R. A., & Sivanesan, S. (2017). Oxidative Stress and Genotoxicity of Zinc Oxide Nanoparticles to Pseudomonas Species, Human Promyelocytic Leukemic (HL-60), and Blood Cells. *Biological Trace Element Research*, *178*(2), 218–227. https://doi.org/10.1007/s12011-016-0921-y.

Tallec, K., Blard, O., González-Fernández, C., Brotons, G., Berchel, M., Soudant, P., Huvet, A., & Paul-Pont, I. (2019). Surface functionalization determines behavior of nanoplastic solutions in model aquatic environments. *Chemosphere*, *225*, 639–646. https://doi.org/10.1016/j.chemosphere.2019.03.077.

Thongkam, W., Gerloff, K., van Berlo, D., Albrecht, C., & Schins, R. P. F. (2017). Oxidant generation, DNA damage and cytotoxicity by a panel of engineered nanomaterials in three different human epithelial cell lines. *Mutagenesis*, *32*(1), 105–115. https://doi.org/10.1093/mutage/gew056.

Ursini, C. L., Cavallo, D., Fresegna, A. M., Ciervo, A., Maiello, R., Tassone, P., Buresti, G., Casciardi, S., Iavicoli, S. (2014). Evaluation of cytotoxic, genotoxic and inflammatory response in human alveolar and bronchial epithelial cells exposed to titanium dioxide nanoparticles. *Journal of Applied Toxicology: JAT*, *34*(11), 1209–1219. https://doi.org/10.1002/jat.3038.

Vales, G., Rubio, L., & Marcos, R. (2015). Long-term exposures to low doses of titaniumx dioxide nanoparticles induce cell transformation, but not genotoxic damage in BEAS-2B cells. *Nanotoxicology*, *9*(5), 568–578. https://doi.org/10.3109/17435390.2014.957252.

Vales, G., Rubio, L., & Marcos, R. (2016). Genotoxic and cell-transformation effects of multi-walled carbon nanotubes (MWCNT) following in vitro sub-chronic exposures. *Journal of Hazardous Materials*, *306*, 193–202. https://doi.org/10.1016/j.jhazmat.2015.12.021.

Vance, M. E., Kuiken, T., Vejerano, E. P., McGinnis, S. P., Hochella, M. F., Rejeski, D., & Hull, M. S. (2015). Nanotechnology in the real world: Redeveloping the nanomaterial consumer products inventory. *Beilstein Journal of Nanotechnology*, *6*(1), 1769–1780. https://doi.org/10.3762/bjnano.6.181.

Vila, L., García-Rodríguez, A., Cortés, C., Marcos, R., & Hernández, A. (2018). Assessing the effects of silver nanoparticles on monolayers of differentiated Caco-2 cells, as a model of intestinal barrier. *Food and Chemical Toxicology*, *116*(Pt B), 1–10. https://doi.org/10.1016/j.fct.2018.04.008.

Wang, J., Che, B., Zhang, L. W., Dong, G., Luo, Q., & Xin, L. (2017). Comparative genotoxicity of silver nanoparticles in human liver HepG2 and lung epithelial A549 cells. *Journal of Applied Toxicology*, *37*(4), 495–501. https://doi.org/10.1002/jat.3385.

Wang, X., Li, T., Su, X., Li, J., Li, W., Gan, J., Wu, T., Kong, L., Zhang, T., Tang, M., & Xue, Y. (2019). Genotoxic effects of silver nanoparticles with/without coating in human liver HepG2 cells and in mice. *Journal of Applied Toxicology*, *39*(6), 908–918. https://doi.org/10.1002/jat.3779.

Wąsowicz, W., Cieślak, M., Palus, J., Stańczyk, M., Dziubałtowska, E., Stępnik, M., & Düchler, M. (2011). Evaluation of biological effects of nanomaterials. Part I. Cyto- and genotoxicity of nanosilver composites applied in textile technologies. *International Journal of Occupational Medicine and Environmental Health*, *24*(4). https://doi.org/10.2478/s13382-011-0041-z.

Watson, C., Ge, J., Cohen, J., Pyrgiotakis, G., Engelward, B. P., & Demokritou, P. (2014). High-Throughput Screening Platform for Engineered Nanoparticle-Mediated Genotoxicity Using CometChip Technology. *ACS Nano*, *8*(3), 2118–2133. https://doi.org/10.1021/nn404871p.

Weir, A., Westerhoff, P., Fabricius, L., Hristovski, K., & von Goetz, N. (2012). Titanium Dioxide Nanoparticles in Food and Personal Care Products. *Environmental Science & Technology*, *46*(4), 2242–2250. https://doi.org/10.1021/es204168d.

Yazdimamaghani, M., Moos, P. J., Dobrovolskaia, M. A., & Ghandehari, H. (2019). Genotoxicity of amorphous silica nanoparticles: Status and prospects. *Nanomedicine: Nanotechnology, Biology and Medicine, 16*, 106–125. https://doi.org/10.1016/j.nano.2018.11.013.

Zhang, Y., Petibone, D., Xu, Y., Mahmood, M., Karmakar, A., Casciano, D., Ali, S., & Biris, A. S. (2014). Toxicity and efficacy of carbon nanotubes and graphene: the utility of carbon-based nanoparticles in nanomedicine. *Drug Metabolism Reviews, 46*(2), 232–246. https://doi.org/10.3109/03602532.2014.883406.

Zhou, F., Liao, F., Chen, L., Liu, Y., Wang, W., & Feng, S. (2019). The size-dependent genotoxicity and oxidative stress of silica nanoparticles on endothelial cells. *Environmental Science and Pollution Research, 26*(2), 1911–1920. https://doi.org/10.1007/s11356-018-3695-2.

Zijno, A., De Angelis, I., De Berardis, B., Andreoli, C., Russo, M. T., Pietraforte, D., Scorza, G., Degan, P., Ponti, J., Rossi, F., & Barone, F. (2015). Different mechanisms are involved in oxidative DNA damage and genotoxicity induction by ZnO and TiO2 nanoparticles in human colon carcinoma cells. *Toxicology in Vitro : An International Journal Published in Association with BIBRA, 29*(7), 1503–1512. https://doi.org/10.1016/j.tiv.2015.06.009.

In: A Closer Look at the Comet Assay
Editor: Keith H. Harmon

ISBN: 978-1-53611-028-9
© 2019 Nova Science Publishers, Inc.

Chapter 4

KINETIC APPROACH IN COMET ASSAY: AN OPPORTUNITY TO INVESTIGATE DNA LOOPS

Katerina Afanasieva[*] *and Andrei Sivolob*
Department of General and Medical Genetics,
Taras Shevchenko National University, Kyiv, Ukraine

ABSTRACT

The comet assay is thought to be a sensitive and simple technique to assess DNA single- and double-strand breaks at the level of individual cells. The approach is based on an analysis of parameters of the electrophoretic track (the comet tail) formed during electrophoresis of nucleoids obtained after cell lysis in agarose on a surface of microscopic slide. The physical principles of DNA migration in this electrophoretic system remained to be rather elusive for a long time. To shed light on the mechanisms of the DNA track formation we applied an original approach based on the kinetic measurements in the comet assay. This chapter is focused on our recent results, which allows us to argue that in neutral conditions at low levels of DNA damages (and also in the case of undamaged cells) the comet tail is formed by extended DNA loops and,

---

[*]Corresponding Author's Email: aphon@ukr.net.

more important, these loops are about the same as chromatin loops in the cell nuclei. Our approach gives an opportunity to investigate several parameters of the loops including the loop supercoiling and large-scale features of the loop domain organization (and re-organization) in nucleoids obtained from cells of different types.

Keywords: comet assay, kinetics, DNA loops, chromatin loops

INTRODUCTION

The comet assay (single-cell gel electrophoresis) was first proposed by Östling and Johanson in 1984 as a method for detection of DNA damages in single cells [1]. The assay was based on the study of electrophoretic migration of DNA from so called nucleoids, structures obtained through cell lysis at high ionic strength, immobilized in a thin layer of agarose on microscopic slides. The principle of the evaluation of damage level was simple: more DNA breaks in the nucleoid, more easily the electrophoretic track is formed. The name of the method is related to the appearance of nucleoids after electrophoresis and staining with fluorescent dyes: they resemble comets with tails formed by DNA, which migrates, and heads containing DNA, which cannot do so. There are two main versions of the comet assay: at neutral and alkaline pH [1–7].

The question about mechanisms of the tail formation was extensively discussed from the very beginning of the development of this beautiful approach. The first interpretation was based on earlier works where it has been shown that the nucleoids obtained through the cell lysis contain supercoiled DNA loops attached to some proteins, which are insensitive to lysis conditions [8, 9]. It has been suggested that at neutral pH the comet tails contain extended DNA loops, and such extension is possible (or at least significantly facilitated) provided that the loops are relaxed due to single-strand breaks [1, 10, 11]. An accumulation of the single-strand breaks leads to double strand breaks, and linear DNA fragments may be present in the tails. Another popular interpretation postulates migration of such fragments to be the main mechanism of the tail formation [2, 7, 11,

12]. The fragments are obviously double or single stranded depending on, respectively, neutral or alkaline (pH > 13) conditions. Independently on the mechanism of the tail formation, the well established fact is that DNA exit into the tail is significantly facilitated when the number of DNA breaks increases [3–7, 13–17].

However, an important question about possibility of migration of intact supercoiled loops was not addressed for a long time. Is it possible to investigate loop domain organization in nucleoids using the comet assay? If yes, then which parameters of this organization can be detected? Do the nuleoid loops correspond, and at which extent, to the chromatin loops in living cell? Here we will focus on our recent experiments, which can give answers to these questions. It should be noted that our analysis will be confined to the comet assay at neutral pH only, because it is hard to say about loop structure of nucleoids in conditions of total DNA denaturation.

KINETIC APPROACH IN THE COMET ASSAY

To analyze the process of DNA migration in the comet assay, we have used a kinetic approach [18–20]. To measure the kinetics of comet formation, several slides, which were simultaneously prepared in the same way, were placed into an electrophoresis tank, and then they were taken out at different time of electrophoresis for staining with a fluorescent dye and analysis. The time dependences of two parameters were analyzed: the relative DNA amount in the tails and the tail length. Conventionally, the first parameter was defined as the ratio of the tail fluorescence intensity to the total intensity of the comet, and the tail length – as the distance from the center of mass of the head to distal end of the tail.

A theory of electrophoresis of a flexible polyelectrolyte trapped in the gel due to a bulky label shows that the time to stretch the chain to its contour length is at most of the order of the reptation time, i.e., very short [21]. It means that the kinetics of DNA amount in the tails, which is rather slow (Figure 1), reflects accumulation of stretched loops rather than the process of loop extension. Therefore, the dependence on time t of the relative amount of DNA in the tails should satisfy the well-known relation

$$\frac{df}{dt} = k(A - f), \qquad (1)$$

where f is the DNA fraction in the tail, k is the rate constant, and A is the maximum relative amount of DNA that can exit. The solution of this equation

$$f = 1 - \exp(-kt) \qquad (2)$$

describes well the kinetic plot obtained for irradiated cells (Figure 1).

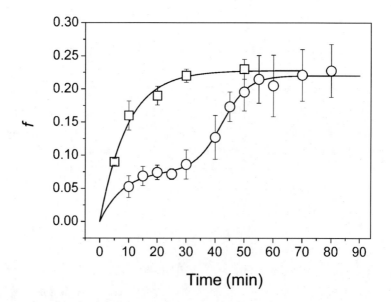

Figure 1. The kinetics of the comet formation. The average relative amount of DNA in the comet tails (*f*) as a function of electrophoresis duration for non-irradiated nucleoids (○) and nucleoids irradiated at 0.5 Gy (□). Continuous curves were obtained by fitting with Eqs. (2) and (4) as described in the text. (Adapted from [18, 19]).

However, the kinetics of DNA exit from nucleoids obtained from undamaged (non-irradiated) cells appeared to be more complex (Figure 1): the plot had a two-step shape. While the first rapid step is also described by Eq. (2), the second delayed step is sigmoid. This reflects some cooperativity of DNA exit at this stage when the rate is increased during the migration. The most attractive explanation of this cooperativity (we will return in the next section to discuss it) is that DNA inside the nucleoid creates friction, which should be decreased during DNA migration into the tail. In other words, the rate of migration should depend not only on the fraction of DNA remained in the head but also on the fraction of DNA that is already came into the tail. Therefore, in the simplest form, the rate equation for the second delayed step can be written as [19]

$$\frac{df}{dt} = k(A-f)\frac{f}{A}, \tag{3}$$

where f is the DNA fraction in the tail, k is the rate constant, A is the maximum relative amount of DNA that can exit in the cooperative regime. The solution of this equation is the well-known Boltzmann equation:

$$f = 1/[1+\exp(k(t_0 - t))], \tag{4}$$

where t_0 is the transition half-time. Figure 1 clearly shows that Eq. (4) fits very well to the second step of the kinetic plot.

THE COMET TAILS MAY CONSIST OF INTACT LOOPS

We have used the simple formalism described in the previous section to analyze the kinetics of the comet formation [18, 19]. Two observations were important. First, the relative amount of DNA in the tails was quite large reaching ~22% after about an hour of electrophoresis. Second, the migration was observed to occur in two steps: some small DNA fractionexited into the tail very rapidly, a larger part of DNA migrated at the second lagging sigmoid step, and then the relative amount of DNA in the comet tails remained constant (Figure 1). Since the nucleoids were intact, a logic assumption could be that the comet tail is formed in this case by intact DNA loops attached with their ends to residual protein structures.

This assumption was confirmed in several experiments. First of all, in agreement with many other observations [4, 7, 12–17], the comet tail formation was accelerated when DNA breaks were introduced in nucleoids by X-rays [18]. The two-step shape of the kinetic plot disappeared in this case, and the maximum DNA amount in the tails was observed at the first step of migration (Figure 1). The acceleration was achieved at low radiation dose (0.5 Gy), which certainly cannot induce DNA fragmentation but can provide relaxation of the loops due to single-strand breaks. The comparison of the kinetic plots obtained for irradiated and non-irradiated nucleoids allowed us to conclude that (i) DNA of non-irradiated cells is mostly intact; and (ii) DNA that migrates at the first rapid step from non-irradiated nucleoids may contain, however, some relaxed nicked loops.

The second evidence for intact supercoiled loops to be the main component of the comet tails of undamaged cells came from the experiments on the reversibility of DNA migration from the comet head [18, 19]. The reasoning to perform such experiments was the following. When intact supercoiled loops are extended during electrophoresis they have to overcome, except agarose resistance, their own torsional constraint: the negative torsional deformations should appear in the loops when they are stretched by the electric force. It can be obviously assumed that disappearance of the external force should cause a reverse contraction of the loops: like springs, they have to go back under the elastic forces. In agreement with the expectation, a gradual decrease in the DNA amount in the tails was observed when the current was switched off after a long-term electrophoresis (Figure 2). After about two hours incubation the relative amount of DNA in the tails reached the level of ~0.1, i.e., approximately the value at the first plateau of the initial (onward) kinetic curve shown in Figure 1. Thus, the second retarded step of the onward migration corresponds to intact supercoiled loops.

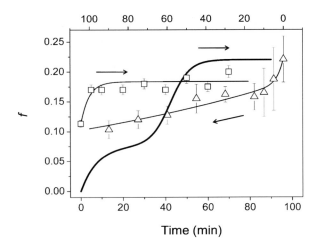

Figure 2. The reversibility of the DNA exit. The average relative amount of DNA in the comet tails (*f*) as a function of electrophoresis duration (bottom abscissa) or time of incubation in the electrophoresis buffer after switching off prolonged electrophoresis (top abscissa). The direction is indicated by arrows near the curves: first onward curve (the smooth curve from Figure 1), backward curve (Δ), and second onward curve (□). (Adapted from [19]).

As it was noted above, this retarded step appeared to be sigmoid, possibly reflecting a cooperativity of the process. However, when the current was reset after a long-term incubation in the electrophoretic buffer, a hysteretic behavior was observed [19]: DNA exited into the tail without the delay, up to the maximum level (Figure 2). This observation allowed us to suggest that the cooperativity of the first onward migration is related to a friction created by DNA inside the nucleoid: a decrease in the DNA concentration inside the comet head due to DNA migration into the tail facilitates the migration of the following loops (Eq. (3)).

This explanation of the cooperativity was supported by our experiments with nucleoids obtained from isolated nuclei [22]. In this case the two-step shape of the kinetic curve disappeared. This effect is probably due to agarose penetration into the nuclei, but not into cells, before polymerization. Therefore, for the nuclei-derived nucleoids, the medium inside the comet head is made by both agarose and DNA, and agarose creates a constant friction. In contrast, inside the cell-derived nucleoids the friction, which is made by DNA only, changes during electrophoresis.

Returning to the loops of the cell-derived nucleoids, which were contracted after switching off electrophoresis: the contraction under elastic forces should not bring the loops to enter inside the nucleoid where the DNA concentration remains high. So, when the current is reset, the loops that are now located on the nucleoid surface migrate rapidly, without the additional friction. Therefore, two conclusions can be made: (i) the second retarded cooperative step of the initial kinetic plot can be attributed to the supercoiled loops inside the nucleoid; (ii) the first rapid step corresponds to loops on the nucleoid surface, which are not necessarily nicked. Since the DNA amount in the tails at this first step is not sensitive to intercalators and protein denaturants (see below), the DNA migration at this step may be just a result of looping from the nucleoid surface. In fact, the kinetic approach in the comet assay makes it possible to distinguish between three kinds of the loops: surface loops; supercoiled loops inside the nucleoid; and loops inside the nucleoid that are too large to exit during electrophoresis. It should be mentioned that the DNA amount in the loops of the three kinds depends on the functional state of cells and the cell type (see below).

The third direct evidence for the inner loops to be supercoiled (and thus intact) came from our experiments on the effects of DNA intercalators on the migration efficiency [18, 19]. The intercalation is associated with an unwinding of the double helix in the intercalation site. Provided that the loop is topologically constrained (i.e., nicks are absent), the unwinding introduces a positive supercoiling. At some binding density of intercalators this positive supercoiling will compensate the initial negative supercoiling, i.e., the loops will be relaxed. It was found, in agreement with this reasoning, that, when the comet assay was performed in the presence of ethidium bromide or chloroquine at concentrations that correspond to the relaxation, a significant acceleration of the DNA migration at the second step occurred (while the first rapid step was not affected). In fact, the second step was shifted and superimposed on the first one. Further increase in the concentration of the intercalators induced a slowdown of the DNA migration. This can be obviously explained by the accumulation of positive supercoiling, which impedes the migration (same as negative supercoiling in the absence of intercalators). The dependence of the migration rate upon the intercalator concentration not only clearly shows that the inner loops are supercoiled, but also can be used to estimate the supercoiling level in the loops (Figure 3). This approach can be used to examine the topological state of DNA in nucleoids prepared from different cells in different functional states. The approach can be also combined with fluorescence *in situ* hybridization (FISH) [23–26] to investigate the topological state of individual loops.

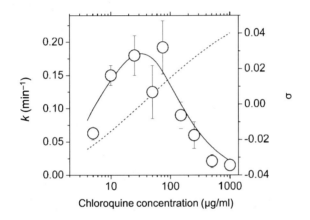

Figure 3. The dependence of the DNA exit on supercoiling: the rate constant k of the DNA exit (points and solid curve) and the net supercoiling density σ in the loops (dashed curve) as functions of chloroquine concentration. The net supercoiling density $\sigma = \sigma_0 + \sigma_i$, where σ_0 is the initial supercoiling density in the nucleoid loops, σ_i is the contribution from intercalation, which depends on the binding density of the intercalator and the unwinding angle in the intercalation sites [19]. The solid curve was obtained by fitting with equation $k=k_0\exp(-\lambda\sigma^2)$, where k_0 is the supercoiling-independent term, λ is the supercoiling force constant [19]. (Adapted from [19]).

LOOPS IN NUCLEOIDS CORRESPOND TO CHROMATIN LOOPS IN NUCLEI

An increase in the concentration of the intercalators was accompanied not only by the slowdown of DNA migration but also by another effect: a decrease in the DNA fraction that can migrate into the tail: the saturation level of the DNA amount in the tails gradually decreased with increasing intercalator concentration and at high concentrations reached the first plateau of the kinetic curve in the absence of the intercalators [18, 19]. Since the comet assay has a limitation with respect to size of the loops that can migrate, it means that the amount of DNA in very large loops (the third kind of the loops mentioned above) increases. This effect can be explained by local unwinding of the double helix in the intercalation sites that can interfere with DNA-protein interactions at the loop ends. The detachment of the ends should obviously give loops of larger sizes, thus decreasing the DNA amount in the loops that can exit. Accordingly, formaldehyde cross-linking of DNA to proteins prevented the decrease in the maximum DNA amount in the comet tails under increase of the concentration of intercalators [19].

The same effect of the decrease in the maximum DNA amount in the comet tails was observed under nucleoid treatments with protein denaturants – urea or sodium dodecyl sulfate [19]. In this case the detachment of the loop ends was obviously achieved due to denaturation of proteins, which were insensitive to the lysis procedure and remained attached before the treatment with denaturants. It should be mentioned that rather gentle denaturation (2M urea), probably on the level of protein–protein interactions, was sufficient to detach the loop ends.

The effects of denaturants and intercalators at high concentrations imply that the anchoring of the loop ends is provided by both protein-DNA and protein-protein interactions. This is in good agreement with the well-known mechanism of the loop anchoring by DNA binding protein CTCF, which, in turn, interacts with cohesin complex [27–32]. This also gives the answer why the loop anchors remain after the cell lysis at high ionic strength. The point is that while CTCF dissociates from DNA in high salt, it probably still interacts with cohesin, which is bound to DNA topologically and cannot be removed by high ionic strength [33, 34]. Then, when the ionic strength is decreased after the lysis procedure, the loop anchors can be reestablished. Therefore, an important conclusion can be made: DNA loops in nucleoids obtained after cell lysis are about the same as chromatin loops in the cell nuclei.

This conclusion was additionally confirmed in experiments with nucleoids obtained in low-salt solution (1 M NaCl) [35]. Usually, and this was the case for all the experiments mentioned above, the cell lysis is performed in high-salt buffer (2.5 M NaCl), which ensures the removal of most chromatin proteins. The nucleoids obtained in 1 M NaCl keep most of histones that are included in a mixture of nucleosome and sub-nucleosome particles: histone octamers (H2A-H2B-H3-H4)$_2$, hexamers H2A-H2B-(H3-H4)$_2$, and tetramers (H3-H4)$_2$ (about half of H2A-H2B dimers are removed in these conditions [36, 37]). It was found that, despite some quantitative differences, the most general features of the kinetics of DNA exit are about the same for nucleoids obtained in high- and low-salt conditions.

The resemblance of the nucleoid DNA loops to chromatin loops gives an interesting opportunity to investigate the loop domain organization of chromatin using relatively simple technique, the comet assay.

THE LOOP SIZE DISTRIBUTION

Except the relative amount of DNA in the tails as a function of electrophoresis duration, we have measured also another parameter, the tail length. The length, which changed usually approximately in parallel with the DNA amount, gives an estimate of the contour length of the largest loops in the tail [19, 38]. Simultaneous measurement of the two parameters, the relative amount f of DNA in the tails and the contour length s_m of the largest loops gives an opportunity to estimate some features of the loop size distribution. The two parameters (at given time of electrophoresis) are obviously related to each other as:

$$f \propto \int_0^{s_m} sP(s)ds, \qquad (5)$$

where s is the loop contour length, and $P(s)$ is the fraction (the probability) of loops of this size. If the kind of the $P(s)$ distribution is known (or can be assumed) then the dependence f vs s_m can be obtained explicitly.

As it was mentioned above, the chromatin loops are known to be anchored by cohesin and CTCF proteins bound to their numerous binding sites (CTCF motifs). Assuming that the motifs are distributed randomly in the genome, the loop contour length distribution $P(s)$ should obey the exponential decay formula [38]:

$$P(s) = \gamma \exp(-\gamma s), \qquad (6)$$

where γ is the average linear loop density in the genome (e.g., the number of loops per kilobase).

The most powerful method to resolve the higher order chromatin structure is Hi-C, the approach which allows mapping of interactions between distant chromatin loci [39–41]. It has been shown, in particular, that an essential part of the genome is organized into loop domains. About 10,000 loops with median contour length 185 kb were detected in human cells [27]. Our analysis of the Hi-C data has shown that the contour lengths of CTCF-cohesin-anchored loops are distributed exponentially, i.e., Eq. (6) fits very well the data. Substituting Eq. (6) into Eq. (5), one can easily get an equation for f as a function of s_m in the explicit form. Figure 4 shows that this equation describes successfully the curves $f$vs$s_m$obtained in our comet assay experiments [38].

Concerning a comparison between the results of Hi-C and the comet assay, several notes have to be made. First, the resolutions of the two methods are very different. In contrast to Hi-C, the comet assay can resolve the loops not longer than ~200 kb. On the other hand, Hi-C can resolve the loops not shorter that 35 kb, which are "visible" in the comet assay. At the same time, our analysis has shown that, according to Hi-C data obtained for GM12875 human lymphoblastoid cells [27], the relative amount of DNA in the loops up to 200 kb is 14.4% (not published), the value which practically coincides with the DNA fraction that exit in the comet assay at the second step from human lymphocyte-derived nucleoids (Figure 1).

However, the values of the loop density γ obtained by the comet assay data analysis (Figure 4) are about one order of magnitude higher than the value obtained from Hi-C data. This is not surprising, taking into account the difference in the resolution between the two methods. In addition, as it was mentioned above, some fraction of DNA (especially the surface DNA exiting at the first step) probably does not correspond to real loops – the first step may represent just DNA pulled out from the nucleoid surface by electric force. On the other hand, much more loops of lower size are seen in the comet assay, including the loops attached to nuclear lamina, for example.

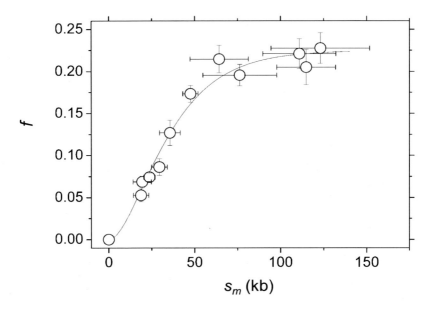

Figure 4. The loop size distribution: correlation between the average relative amount f of DNA in the comet tails and the average contour length of the longest loops in the tail s_m. Continuous curve was obtained by fitting with Eqs. (5) and (6). The loop density γ was found to be 0.53 ± 0.03 kb^{-1}. (Adapted from [36].)

Nevertheless, although the γ value obtained in the comet assay does not directly reflect the real number of loops detected in Hi-C experiments, this value can be taken as a relative measure of the loop density. This can be especially useful for comparisons between different cell types.

LOOP DOMAIN ORGANIZATION IN DIFFERENT CELLS

The chromatin loops, which frequently connect promoters and enhancers are thought to vary between cells with different transcriptional activity [27, 43, 44]. Indeed, we have shown that the loop density estimated in the comet assay (previous section) varies for different cell types. Rather high values of the density of short loops (up to ~150 kb) were found for cells, which are not very active in transcription (human lymphocytes and T98G human glioblastoma cells, which were "frozen" at G1 phase). In contrast, the estimated loop density decreases approximately twice upon lymphocyte activation by interleukin 2 [38]. A reactivation of the T98G cells was also accompanied by the same decrease in the loop density (not published). These observations probably mean that, upon cell activation, the number of relatively short (up to ~150 kb) loops decreases in favor of larger loops, which cannot be resolved in the comet assay.

We observed also another kind of the loop redistributions in our experiments. The two-step behavior of the DNA exit (Figure 1) was found to be a universal characteristic of DNA migration: the same two steps were seen in the kinetic curves obtained for activated lymphocytes (lymphoblasts) and T98G cells. The relative contributions of the steps (i.e., of "rapid" (surface) and "slow" (inner) loops) were, however, different for different cells [38]. For the lymphoblast-derived nucleoids the maximum amplitude of the second component was two-times lower in comparison with lymphocytes while the total amount of DNA in the tails after long-term electrophoresis remained the same (respectively, DNA fraction at the first rapid step was two-times higher). In other words, a redistribution of the loop domains between the inside and surface "fractions" occurs upon lymphocyte activation.

Thus, two parameters, which can be detected in the comet assay, the ratio between DNA amounts in the inner and surface loops, and the density of short (up to ~150 kb) loops, appeared to be sensitive to cell functional transitions and to be dependent on the sell types.

CONCLUSION AND FUTURE PERSPECTIVES

In the comet assay at neutral pH at low levels of DNA damages, and in undamaged cells, the extended DNA loops dominate in the comet tails. An analysis of the kinetics of DNA exit during the comet assay helps to estimate the supercoiling level of these loops, the ratio between DNA amount in the loops inside nucleoids and on its surface, and general features of the loop size distributions. The most important conclusion from the kinetic studies is that the DNA loops in nucleoids obtained after cell lysis are about the same as chromatin loops in the cell nuclei. Respectively, kinetic measurements in the comet assay give an opportunity to investigate the topology of the loops and large-scale features of the loop domain organization (and re-organization) in nucleoids obtained from different cell types.

REFERENCES

[1] Östling, O., and Johanson, K. J. 1984. "Microelectrophoretic study of radiation-induced DNA damages in individual mammalian cells."*Biochem. Biophys. Res. Commun.* 123:291-98.
[2] Singh, N.P. McCoy, M.T., Tice, R.R., and Schneider, E.L. 1988. "A simple technique for quantitation of low levels of DNA damage in individual cells." *Exp. Cell Res.*175:184–91.
[3] Olive, P.L. 2002. "The comet assay. An overview of techniques."*Methods Mol. Biol.*203:179-94.
[4] Collins, A.R. 2004. "The comet assay for DNA damage and repair: principles, applications, and limitations." *Mol. Biotechnol.* 26:249-61.
[5] Hartmann, A., Agurell, E., Beevers, C., Brendler-Schwaab, S., Burlinson, B., Clay, P.,Collins, A., Smith, A., Speit, G., Thybaud, V., and Tice, R.R. 2003. "Recommendations for conducting the in vivo alkaline comet assay." *Mutagenesis* 18:45-51.

[6] Olive, P., andBanáth, J. 2006. "The comet assay: a method to measure DNA damage in individual cells." *Nat. Protoc.* 1:23–9.

[7] Collins, A.R., Oscoz, A.A., Brunborg, G., Gaivão, I., Giovannelli, L., Kruszewski, M., Smith, C.C., and Štětina, R. 2008. "The comet assay: topical issues." *Mutagenesis* 23:143-51.

[8] Cook, P.R., andBrazell, I.A. 1975. "Supercoils in human DNA." *J. Cell Sci.* 19:261–79.

[9] Cook, P.R., Brazell, I.A., andJost, E. 1976. "Characterization of nuclear structures containing superhelical DNA." *J. Cell Sci.* 22:303–24.

[10] Collins, A.R., Dobson, V.L., Dusinská, M., Kennedy, G., andŠtětina, R. 1997. "The comet assay: what can it really tell us?" *Mutat. Res.* 375:183–93.

[11] Shaposhnikov, S.A., Salenko, V.B., Brunborg, G., Nygren, J., and Collins, A.R. 2008. "Single-cell gel electrophoresis (the comet assay): loops or fragments?" *Electrophoresis* 29:3005–12.

[12] Olive, P.L., Wlodek, D., andBanáth, J.P. 1991. "DNA double-strand breaks measured in individual cells subjected to gel electrophoresis." *Cancer Res.* 51:4671–6.

[13] Fairbairn, D.W., Olive, P.L., and O'Neal, K.L. 1995. "The comet assay: a comprehensive review." *Mutat. Res.* 339:37–59.

[14] Brendler-Schwaab, S., Hartmann, A., Pfuhler, S., andSpeit, G. 2005. "The *in vivo* comet assay: use and status in genotoxicity testing." *Mutagenesis* 20:245–54.

[15] Azqueta, A., and Collins, A.R. 2013. "The essential comet assay: a comprehensive guide to measuring DNA damage and repair." *Arch. Toxicol.* 87:949–68.

[16] Glei, M., Schneider, T., andSchlörmann, W.2016. "Comet assay: an essential tool in toxicological research." *Arch. Toxicol.* 90:2315–36.

[17] Møller, P. 2018. "The comet assay: ready for 30 more years." *Mutagenesis* 33:1–7.

[18] Afanasieva, K., Zazhytska,M., andSivolob, A. 2010. "Kinetics of comet formation in single-cell gel electrophoresis: loops and fragments." *Electrophoresis* 31:512–19.

[19] Afanasieva, K., Chopei, M., Zazhytska, M., Vikhreva, M., and Sivolob, A. 2013. "DNA loop domain organization as revealed by single-cell gel electrophoresis." *Biochim. Biophys. Acta* 1833:3237–44.
[20] Afanasieva, K, and Sivolob, A. 2018. "Physical principles and new applications of comet assay." *Biophys Chem.* 238:1–7.
[21] Défontaines, A.D., and Viovy, J.L.1993. "Gel electrophoresis of an end-labeled DNA. I. Dynamics and trapping in constant fields." *Ectrophoresis* 14:8–17.
[22] Afanasieva, K., Chopei, M., and Sivolob, A.2015. "Single nucleus versus single cell gel electrophoresis: kinetics of DNA track formation." *Electrophoresis* 36:973–7.
[23] Santos, S.J., Singh, N.P., and Natarajan, A.T. 1997. "Fluorescence *in situ* hybridization with comets." *Exp. Cell Res.* 232:407–11.
[24] Spivak, G.,Cox, R.A., and Hanawalt, P.C. 2009. "New applications of the comet assay: comet-FISH and transcription-coupled DNA repair." *Mutat. Res.* 681:44–50.
[25] Shaposhnikov, S., El Yamani, N., and Collins, A.R.2015. "Fluorescent *in situ* hybridization on comets: FISH comet." *Methods Mol. Biol.* 1288:363–73.
[26] Shaposhnikov, S., Frengen, E., and Collins, A.R.2009. "Increasing the resolution of the comet assay using fluorescent *in situ* hybridization – a review." *Mutagenesis* 24:383–9.
[27] Rao, S.S., Huntley, M.H., Durand, N.C., Stamenova, E.K., Bochkov, I.D.,Robinson, J.T., Sanborn, A.L., Machol, I., Omer, A.D., Lander, E.S., and Aiden,E.L. 2014. "A 3D map of the human genome at kilobase resolution reveals principles of chromatin looping." *Cell* 159:1665–80.
[28] Sanborn, A.L., Rao, S.S.P., Huang, S.C., Durand, N.C., Huntley, M.H., Jewett, A.I., Bochkov, I.D., Chinnappan, D., Cutkosky, A., Li, J., Geeting, K.P., Gnirke, A., Melnikov, A., McKenna, D., Stamenova, E.K., Lander, E.S., and Aiden, E.L. 2015. "Chromatin extrusion explains key features of loop and domain formation in wild

type and engineered genomes." *Proc. Natl. Acad. Sci. USA* 112: E6456–E6465. doi: 10.1073/pnas.1518552112.

[29] Fudenberg, G., Imakaev, M., Lu, C., Goloborodko, A., Abdennur, N., and Mirny, L.A. 2016. "Formation of chromosomal domains by loop extrusion." *Cell Rep.* 15:2038–49.

[30] Nora, E.P., Goloborodko, A., Valton, A.L., Gibcus, J.H., Uebersohn, A.,Abdennur, N., Dekker, J., Mirny, L.A. and Bruneau, B.G. 2017. "Targeted degradation of CTCF decouples local insulation of chromosome domains from genomic compartmentalization." *Cell* 169:930–44.

[31] Rao, S.S.P., Huang, S.C., Glenn St Hilaire, B., Engreitz, J.M., Perez, E.M., Kieffer-Kwon, K.R., Sanborn, A.L., Johnstone, S.E., Bascom, G.D., Bochkov, I.D., Huang, X., Shamim, M.S., Shin, J., Turner, D., Ye, Z., Omer, A.D., Robinson, J.T., Schlick, T.,Bernstein, B.E., Casellas, R., Lander, E.S., and Aiden, E.L.2017. "Cohesin loss eliminates all loop domains." *Cell* 171:305–20.

[32] Vian, L., Pękowska, A., Rao, S., Kieffer-Kwon, K., Jung S., Baranello, L., Huang, S., El Khattabi, L., Dose, M., Pruett, N., Sanborn, A., Canela, A., Maman, Y., Oksanen, A., Resch, W., Li, X., Lee, B., Kovalchuk, A,, Tang, Z., Nelson, S., Di Pierro, M, Cheng, R., Machol, I., St Hilaire, B., Durand, N., Shamim, M., Stamenova, E., Onuchic, J., Ruan, Y., Nussenzweig, A., Levens, D., Aiden, E. and Casellas, R. 2018. "The energetics and physiological impact of cohesin extrusion." *Cell* 173:1165–78.

[33] Onn, I., and Koshland, D. 2011. "*In vitro* assembly of physiological cohesin/DNA complexes." *Proc. Natl. Acad. Sci. USA* 108: 12198–205.

[34] Stigler, J., Çamdere, G.Ö., Koshland, D.E., and Greene, E.C. 2016. "Single-molecule imaging reveals a collapsed conformational state for DNA-bound cohesin." *Cell Rep.* 15:988–98.

[35] Afanasieva, K.S., Olefirenko, V.V., and Sivolob, A.V. 2018. "DNA loops after cell lysis resemble chromatin loops in an intact nucleus." *Ukr. Biochem. J.* 90: 43–9.

[36] Eickbush, T.H., and Moudrianakis E.N. 1978. "The histone core complex: an octamer assembled by two sets of protein-protein interactions." *Biochemistry* 17:4955–64.
[37] Khrapunov, S.N., Dragan, A.I., Sivolob, A.V., and Zagariya, A.M. 1997. "Mechanisms of stabilizing nucleosome structure. Study of dissociation of histone octamer from DNA." *Biochim. Biophys. Acta.* 1351:213–22.
[38] Afanasieva, K., Chopei, M.,Lozovik, A., Semenova, A., Lukash, L., and Sivolob, A. 2017. "DNA loop domain organization in nucleoids from cells of different types." *Biochem. Biophys. Res. Commun.* 483:142–6.
[39] Lieberman-Aiden, E.,. van Berkum, N.L., Williams, L., Imakaev, M., Ragoczy, T., Telling, A., Amit, I., Lajoie, B.R., Sabo, P.J., Dorschner, M.O., Sandstrom, R., Bernstein, B., Bender, M.A., Groudine, M., Gnirke, A., Stamatoyannopoulos, J., Mirny, L.A., Lander, E.S., and Dekker, J. 2009. "Comprehensive mapping of long-range interactions reveals folding principles of the human genome." *Science* 326:289–93.
[40] Dekker, J., Marti-Renom, M.A., and Mirny, L.A. 2013. "Exploring the three-dimensional organization of genomes: interpreting chromatin interaction data." *Nat. Rev. Genet.* 14:390–403.
[41] Gibcus,J.H., and Dekker, J. 2013."The hierarchy of the 3D genome." *Mol. Cell* 49:773–82.
[42] Sexton, T., and Cavalli, G. 2015. "The role of chromosome domains in shaping the functional genome." *Cell* 160:1049–59.
[43] Tang, Z., Luo, O.J., Li, X., Zheng M, Zhu, J.J., Szalaj, P., Trzaskoma, P., Magalska, A., Wlodarczyk, J., Ruszczycki, B., Michalski, P., Piecuch, E., Wang, P., Wang, D., Tian, S.Z., Penrad-Mobayed, M., Sachs, L.M., Ruan, X., Wei, C.L., Liu, E.T., Wilczynski, G.M., Plewczynski, D., Li, G., and Ruan, Y. 2015. "CTCF-mediated human 3D genome architecture reveals chromatin topology for transcription." *Cell* 163:1611–27.

[44] Ji, X., Dadon, D.B., Powell, B.E., Fan, Z.P., Borges-Rivera, D., Shachar, S., Weintraub, A.S., Hnisz, D., Pegoraro, G., Lee, T.I., Misteli, T., Jaenisch, R., and Young, R.A. 2016. "3D Chromosome regulatory landscape of human pluripotent cells." Cell Stem Cell 18:262–75.

In: A Closer Look at the Comet Assay
Editor: Keith H. Harmon

ISBN: 978-1-53611-028-9
© 2019 Nova Science Publishers, Inc.

Chapter 5

EVALUATION OF GLOBAL DNA METHYLATION STATUS OF SINGLE CELLS BY THE COMET ASSAY: A PROMISING APPROACH IN CANCER DIAGNOSIS AND FOLLOW-UP

Yildiz Dincer[*], PhD

Istanbul University-Cerrahpasa, Cerrahpasa Medical Faculty,
Department of Medical Biochemistry, Istanbul, Turkey

ABSTRACT

As an epigenetic modification, DNA methylation plays a pivotal role in gene regulation, and is crucial for maintaining genome stability. DNA methylation occurs generally in cytosine within CpG dinucleotides forming 5-methylcytosine (5-mC) on gene promoters and is associated with repression of transcription. This epigenetic event is reversible, removal of methyl group causes transcription of the gene. DNA methylation patterns can be established on a global or gene-specific level in accordance with regulatory needs of cells. If the DNA methylation process is not correctly regulated, it can lead to disruption in gene

---

[*] Corresponding Author's Email: yldz.dincer@gmail.com

expression, impairment in biological pathways and finally development of a wide range of diseases. Aberrant DNA methylation is the most common feature of tumor cells. Genomic instability, oncogene activation and tumor suppressor gene inactivation through altered DNA methylation patterns are well defined in tumor cells. DNA methyltransferase inhibitors, which are designed to demethylate DNA, have already been introduced into clinical use for cancer therapy. In order to detect DNA methylation signature in cancer diagnosis and prognosis, various methods have been established. DNA sequencing after bisulfite conversion is the most frequently used assay to measure both global and gene-specific DNA methylation. However, all methods are expensive, require long time and there is no standardization and unification. Comet assay is a method for assessment of DNA damage and repair quantitatively at the level of single cell. The assay has been used in testing genotoxicity of chemicals, monitoring environmental contamination with genotoxic compounds, human biomonitoring and molecular epidemiology. In addition to DNA strand breaks, some base lesions can be measured by the comet assay using lesion-specific endonucleases. In this context, comet assay is frequently used to determine oxidative DNA damage using 8-oxoguanine glycosylase 1. Recently, a new modification of the comet assay was established to determine the global DNA methylation level of individual cells using 5-methylcytosine (5-mC)-specific restriction endonucleases that are HpaII, MspI and McrBC. The EpiComet-Chip, a modified high-throughput comet assay for the determination of DNA methylation status was introduced. In this chapter, this new application of the comet assay is decribed for detection of aberrant DNA methylation which is a promising marker in cancer diagnosis and follow-up.

Keywords: DNA methylation, cancer, comet assay, EpiComet, EpiComet-Chip

INTRODUCTION

Recent technical developments and their transfer into area of medical research caused a rapid progress in diagnosis and treatment of human diseases. Less than a century ago, very little was known about the role of genetic factors in human disease. After discovery of DNA structure by James Watson and Francis Crick in 1953, methods to determine the base sequences in DNA were developed in the mid 1970s. The *Human Genome*

Project started to complete mapping and understanding of all the genes of *human* beings in 1990. The *Human Genome Project* finished in 2003 and provided researchers powerful knowledge to understand the genetic base of human diseases, paving the way for new strategies for their diagnosis, treatment and prevention. Today, researchers can find a gene suspected of causing an inherited disease with a genetic test in a short time [1]. However genetic changes, in the term of mutation, can not always enough to explain underlying mechanism of diseases. Whilst the science of genetics was making rapid progress, researchers were focused on regulation of gene expression. Discovery of epigenetic mechanisms opened a new era in this field. The mechanisms, which explain how expression of a gene is stopped or how a gene is overexpressed, although there is no change in its base sequence, provided a better understanding of pathogenesis of diseases. Epigenetic changes are affected by environmental factors and are reversible. This feature allowed the development of a new therapeutic approach, epigenetic therapy.

EPIGENETIC MECHANISMS

Epigenetic refers to external modifications to DNA that turn genes "on" or "off" without affecting the DNA base sequence. Epigenetic modifications are typically divided into three categories: DNA methylation, histone post-translational modifications and microRNA-mediated modifications. Epigenetic modifications are acquired throughout life depend on environmental clues such as diet, lifestyle and toxin exposure, and are heritable. Although all cells in one individual have the same DNA sequence, epigenetic regulation occurs at the specific gene loci in the specific cells to yield specific cellular phenotypes. A change in the phenotype does not usually effect the genotype.

DNA methylation occurs as covalent addition of a methyl group to the 5-carbon of cytosine forming 5-methylcytosine (5-mC) mostly located at cytosinephosphate-guanine sites (CpG) that are present in the 5'-untranslated regions of gene promoters. DNA methylation is catalyzed by

DNA methyltransferases (DNMTs), and methyl group donor is S-adenosyl-L-methionine (SAM). Mammalian DNA contains two methylated cytosine bases; 5-methylcytosine (5mC) and 5-hydroxymethylcytosine (5hmC). Ten-Eleven-Translocation (TET) family of oxygenases catalyzes the conversion of 5mC to 5hmC. Transcription level of a gene is affected by 5mC and 5hmC at the promoter region via two ways: methylated DNA prevents the binding of transcription factors to the gene promoter and methylated DNAs are occupied by methyl-CpG-binding domain proteins (MeCPs). MeCPs bring together the other epigenetic components, form compact and inactive heterochromatin, and consequently cause gene silencing. On the contrary, hypomethylation results in activation of transcription. DNA methylation patterns can be established on a global or gene-specific level in accordance with regulatory needs.

As the second epigenetic mechanism, chromatin can be exposed to modifications. These modifications may cause changes in chromatin structure which in turn effect the expression patterns of the genes. Modifications of histone proteins alter the access of the transcriptional machinery to genes in DNA. Acetylation, phosphorylation, methylation, glycosylation and ubiquitination are the modifications which can modulate biological activity of histones.

The third epigenetic regulatory mechanism is maintained by non-coding microRNAs. Briefly, they bind to their target mRNAs at their 3'-untranslated regions, supress their translation and/or promote their degradation, and finally repress protein synthesis.

DNA METHYLATION SIGNATURE IN CANCER

If the epigenetic processes are not correctly regulated, it may lead to changes in DNA methylation and histone modification patterns that disrupt gene expression and cellular processes including DNA repair and tumor suppression. Maintaining genome stability is crucial for the prevention of tumor formation. DNA methylation not only plays an important role in regulation of gene expression but also is essential for maintaining genome stability. Aberrant DNA methylation is a common feature of human tumors. It is an early event in tumor development and is considered as a driving mechanism of carcinogenesis. Certain biological processes, exogenous chemicals such as pharmaceuticals and food additives cause aberrant DNA methylation by changing DNA methylation patterns [2-5]. Aberrant DNA methylation referes to hypomethylation and hypermethylation. Hypomethylation occurs in both CpG dinucleotides in repetitive elements and gene-specific promoter CpG islands, whereas hypermethylation is often observed at CpG islands in gene promoters. Genome-wide hypomethylation resulting from demethylation in repeats or transposable elements or across the genome is associated with genomic instability, increased mutation rate and cancer development [6]. Alterations in DNA methylation pattern at CpG islands in gene promoters may cause aberrant expression of genes that are related to cell proliferation and differentiation, which in turn results in loss of regulation of cell differentiation and transformation, and eventually, tumor formation. Hypomethylation in promoter regions of oncogenes results in their upregulation and overproduction of the oncoproteins. Hypomethylation of the promoter regions of tumor-suppressor genes allow their expression and maintains the normal state of the cell. On the contrary, hypermethylation at promoter regions of tumor suppressor genes is an important mechanism in their inactivation. Promoter regions of tumor suppressor genes are generally hypomethylated in normal cells whereas they are hypermethylated in cancer cells resulting transcriptional silencing (Figure 1). It was detected that not only tumor suppressor genes but also cell cycle control genes, DNA repair genes and pro-apoptotic genes are silenced through promoter hypermethylation in tumor cells. DNA methylation changes are also exist outside of the tumor. Due to high turnover rate, tumor cells release fragments of their DNA into the bloodstream. The analysis of cell-free circulating DNA in the blood is termed liquid biopsy. Liquid biopsy allows for detection of tumor DNA in different aspects such as quantity, mutations and epigenetic changes. Aberrant DNA methylation was detected in DNA obtained from serum, urine, sputum and other body fluids of cancer patients [7-10].

DNA methylation signature opened a new era in diagnosis, prognosis and treatment of cancer. As respect with the fact that DNA methylation changes is an early event in carcinogenesis and are exist outside of the tumor, stable alterations in DNA methylation can serve as a non-invasive biomarker for early detection of cancer and screening in asymptomatic individuals with high risk. On the other hand, in general, first-line therapy is based on a biopsy of the primary tumor. The major obstacle in testing the efficacy of treatment is the limited access to post-treatment tumor tissue. Methylation testing in DNA obtained from serum and/or urine can open a new door in monitoring the treatment. Furthermore in the first-line therapy based on a biopsy of the primary tumor, relevant changes in the metastasis are missed, thus primary resistance occurs. After switching to a second-line therapy, methylating changes of resistant clones can be analyzed in serum and/or urine and resistance mechanisms might be detected before progression [11].

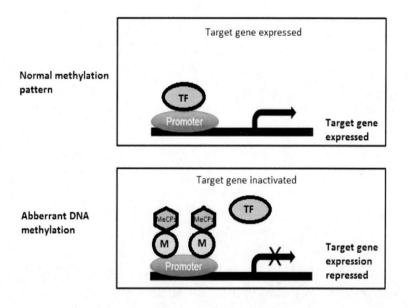

MeCPs: Methyl-CpG-binding domain proteins MeCPs, TF: Transcription factor, M: Methyl groups

Figure 1. Normal and aberrant DNA methylation

METHODS TO DETECT DNA METHYLATION STATUS

Several technologies have been developed to analyse DNA methylation from locus-specific through the entire genome. DNA is a stable and robust molecule so that DNA methylation can be reliably analyzed not only in fresh but also in frozen and archived clinical specimens. Although immunoassays and high-performance liquid chromatography methods are available, DNA sequencing after bisulfite conversion is regarded as a gold-standard technology to measure both global and gene-specific DNA methylation [12]. Pricipally, bisulfite treatment converts unmethylated cytosines into uracil and uracil is recognized as thymine in subsequent PCR amplification and sequencing. However, 5mCs are resistant to this convertion and remain as cytosines allowing 5mCs to be distinguished from unmethylated DNA [13]. Several techniques have arisen based on the working basis of bisulfite convertion including Methylation Specific PCR (MSP), Combined Bisulfite Restriction Analysis (COBRA), Methylation-sensitive Single Nucleotide Primer Extension (Ms-SNuPE) and several other techniques depending on different applications [14-16]. Recently, Shen et al. [17] reviewed all methods used for DNA methylation analysis. Commercial products analyzing genome wide methylation are available. Although high sensitivity and specificity have been obtained in a few studies, overall results are far from satisfactory. This problem likely arised from lack of consensus regarding pre-analitical and technical procedures. Methylation detection approach have varied widely among studies. No one method of DNA methylation analysis will be appropriate for every application. Most importantly, these methods for DNA methylation analysis are difficult, time consuming and expensive.

COMET ASSAY

The comet assay (single cell gel electrophoresis) is a rapid, *simple,* sensitive, versatile *and* cheap method to measure DNA damage *in vivo* and *in vitro*. Comet assay is widely used in various research areas including human biomonitoring, genotoxicity assessment due to environmental and occupational exposure to harmful chemicals, and in studies of DNA damage and DNA repair [18]. A variety of DNA lesions can be detected using comet assay, including DNA double- and single-strand breaks (DSBs and SSBs, respectively) and alkali-labile sites [19]. For the traditional comet assay, after embedding cells in a layer of agarose on a microscope slide they are lysed with detergent and high-salt solution, leaving "nucleoids" containing intact supercoiled loops of DNA. Supercoiling is relaxed by placing the microscope slide in alkali. Damaged DNA extends towards the anode under electrophoresis giving an appearance of the tail of a comet when stained and viewed under fluorescence microscopy. Undamaged DNA remains within the comet head. The relative amount of DNA in the comet tail indicates DNA break frequency [20]. The comet assay can be modified to detect various base damages as DNA single strand breaks by digestion of nucleoids with a lesion-specific endonuclease after lysis step [21]. Damage-specific endonucleases create strand breaks at damaged bases. The most frequently used endonucleases are fpg (formamidopyrimidine DNA glycosylase) breaks at 8-oxoguanine (8-oxoGua) and methyl-fapy-guanine, and Endonuclease III which creates breaks mainly at oxidized pyrimidines.

MODIFIED COMET ASSAY TO DETECT GLOBAL DNA METHYLATION LEVEL

A novel approach including the comet assay is to detect global DNA methylation levels using a methylation-specific restriction endonuclease. Three methylation-specific endonucleases have been introduced for this purpose. Firstly, the isoschizomeric restriction enzymes HpaII and MspI were used. HpaII and MspI recognize the same tetranucleotide sequence (5'-CCGG-3') but have differential sensitivity to DNA methylation. HpaII is inactive when any of the two cytosines is methylated, but it digests the hemimethylated 5'-CCGG-3' at a lower rate compared with the unmethylated sequences. MspI digests 5'-CmCGG-3' but not 5'-mCCGG-3' [22]. This new modification of the comet assay was applied in several biomonitoring studies with combined usage of HpaII and MspI [23-25]. Similar to the common limitations of the other modified comet assays, limited sample throughput, insufficient enzyme digestion of nucleoids and drying of agarose before enzyme digestion were reported as the challenges and limitations of the methylation sensitive comet assay [26].

Recently, McrBC was introduced as a more effective methylation specific restriction enzyme [27]. McrBC cleaves DNA containing 5-methylcytosine, 5-hydroxymethylcytosine or N4-methylcytosine on one or both strands [22]. McrBC recognizes two half-sites on DNA of the form (G/A)mC; these half-sites can be separated by up to 3 kb, with an optimal separation is 55–103 bp (recognition site-5'…PumC (Prurine methylscytosine) (N-40-3000) PumC…3'). This short consensus sequence of McrBC, (PumC) allows the enzyme to recognize and cut a large proportion of the methylcytosines present in DNA [28, 29].

Very recently, Towsend et al. [30] established the 'EpiComet' using McrBC as methylation specific restriction enzyme to detect global DNA methylation status. For the EpiComet assay, the standard procedure of the comet assay is followed. After the lysis, the standard procedure is modified to allow for the determination of global DNA methylation status. Briefly, slides are washed in Wash Buffer and allowed to equilibrate with Wash Buffer at room temperature. Subsequently, samples are incubated at 37°C in a preheated damp chamber by layering either Control Treatment Buffer which includes Wash Buffer plus BSA and GTP (required for McrBC enzymatic activity), or Methylation-Specific Buffer which includes Control Treatment Buffer plus McrBC. The slides are then transferred into a chilled alkaline solution and allowed to remain in the solution to unwind DNA. After unwinding, electrophoresis is performed. The slides and plates are then washed with neutralizing buffer to neutralize the remaining alkali and remove detergent, and dried with ice cold 100% ethanol. Slides are stained, comets are scored and analyzed by the methods described for traditional comet assay (Figure 2). Incubation with McrBC induces de novo DNA strand breaks at a majority of the 5-mC present in the DNA, thus converts undetectable 5-mC into single strand breaks that can be easily quantified by the comet assay. Global DNA methylation is interpreted by measuring and subtracting the mean % Tail DNA from buffer-only slides, which represent basal strand breaks, from the mean % Tail DNA obtained after incubation with the methylation sensitive restriction enzyme. An increase in the % Tail DNA after enzyme treatment is indicative of DNA hypermethylation [30].

The comet assay is simple and economical in the terms of material and equipment but it is labour-intensive. Because there is a limit to the number of samples that can be processed in one experiment. A higher-throughput platform for the comet assay, CometChip (a 96-well platform), has been developed to solve this problem [31-33]. For the CometChip, cells are arrayed in microwells within agarose, and then analyzed for single strand breaks. The CometChip approach uses the same parameters as the traditional comet assay. Key advantages of the CometChip are the higher throughput and increased sensitivity. After establishment of EpiComet, Towsend et al. [30] generated EpiComet-Chip by combining the EpiComet and CometChip technology. This merged approach enables simultaneous analysis of DNA damage and global methylation levels with unprecedented speed and simplicity. They evaluated EpiComet-Chip's potential in predicting exposure-mediated genotoxicity and global DNA methylation alterations at the cellular level in several experiments. They reported that the results obtained using EpiComet-Chip are not different from results versus direct comparison to EpiComet; EpiComet-Chip is capable of correctly detecting global methylation alterations in response to *in vitro* exposures to hypomethylating and hypermethylating agents.

FUTURE PERSPECTIVE

Continued rapid improvements in the field of molecular biology made the study of DNA methylation more accessible. The recent progress in cancer diagnosis and prognosis by DNA methylation signature has required a sensitive, fast and cheap method to measure DNA methylation status of cells isolated from any tissue. Despite several methods were developed, no simple, cost effective and standardized method is available to detect global DNA methylation status using a single platform. EpiComet and EpiComet-Chip are easy, rapid and cheap assays in comparison to other methods that are used to detect global DNA methylation. In addition, using a labelled probe, a sequence of interest can be specifically detected on comets by fluorescence *in situ* hybridisation (FISH). FISH-comet assay can be used for validation of methylation changes identified in EpiComet-Chip experiments. These novel promising technologies may help screening and early detection of cancer as well as monitoring the cancer therapy. However, in order to introduce to clinical use, a method should have high sensitivity and specificity; and inter-and intra-assay reproducibility should be tested. Both EpiComet and EpiComet-Chip are very new and there is no accumulated data yet. Future studies will reveal usefulness of this modified comet assay in early diagnosis and treatment of cancer.

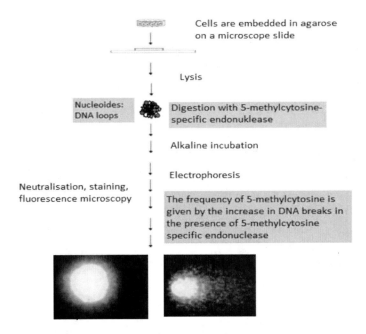

Figure 2. Comet assay modified with 5-methylcytosine specific restriction endonucleases

REFERENCES

[1] "Human Genome Project," National Institutes of Health, Updated October 2010, https://report.nih.gov/nIHfactsheets/Pdfs/HumanGenomeProject(NHGRI).pdf.

[2] Reik, Wolf. (2007). "Stability and flexibility of epigenetic gene regulation in mammalian development." *Nature*, 447(7143), 425–432. doi:10.1038/nature05918.

[3] Csokaa, Antonei B. & Szyfb, Moshe. (2009). "Epigenetic side-effects of common pharmaceuticals: a potential new field in medicine and pharmacology." *Medical Hypotheses*, 73(5), 770–780. doi:10.1016/j.mehy.2008.10.039.

[4] Feil, Robert. & Mario, F. Fraga. (2011). "Epigenetics and the environment: emerging patterns and implications." *Nature Reviw Genetics*, 13(2), 97–109. doi:10.1038/nrg3142.

[5] Benayoun, Bérénice A., Elizabeth, A. Pollina. & Anne, Brunet. (2015). "Epigenetic regulation of ageing: linking environmental inputs to genomic stability." *Nature Revews Molecular Cell Biology*, 16(10), 593–610. doi:10.1038/nrm4048.

[6] Cho, Yoon H., Yoonhee, Jang., Hae, Dong Woo., Yang, Jee Kim., Su Young, Kim., Sonja, Christensen., Elizabeth, Cole., Soo, Yong Choi. & Hai Won, Chung. (2019). "LINE1 Hypomethylation is Associated With Radiation-Induced Genomic Instability in Industrial Radiographers." *Environmental and Molecular Mutagenesis*, 60, 174-84. doi:10.1002/em.22237.

[7] Hoque, Mohammad Obaidul., Shahnaz, Begum., Ozlem, Topaloglu., Carmen, Jeronimo., Elizabeth, Mambo., William, H. Westra., Joseph, A. Califano. & David, Sidransky. (2004). "Quantitative detection of promoter hypermethylation of multiple genes in the tumor, urine, and serum DNA of patients with renal cancer." *Cancer Research*, 64, 5511-7. doi:10.1158/0008-5472.CAN-04-0799.

[8] Dulaimi, Essel., Robert, G. Uzzo., Richard, E. Greenberg., Tahseen, Al-Saleem. & Paul, Cairns. (2004). "Detection of bladder cancer in urine by a tumor suppressor gene hypermethylation panel."

Clinical Cancer Research, *10*, 1887-93. doi:10.1158/1078-0432. CCR-03-0127.

[9] Topaloglu, Ozlem., Mohammad, Obaidul Hoque., Yutaka, Tokumaru., Juna, Lee., Edward, Ratovitski., David, Sidransky. & Chul-so, Moon. (2004). "Detection of promoter hypermethylation of multiple genes in the tumor and bronchoalveolar lavage of patients with lung cancer." *Clinical Cancer Research*, *10*, 2284-8. doi:10.1158/1078-0432.CCR-1111-3.

[10] Huang, Zhao-Hui., Li-Hua, Li., Fan, Yang. & Jin-Fu, Wang. (2007). "Detection of aberrant methylation in fecal DNA as a molecular screening tool for colorectal cancer and precancerous lesions." *World Journal of Gastroenterology*, *13*, 950-4. doi:10.3748/wjg.v13.i6.950.

[11] Heitzer, Ellen., Peter, Ulz. & Jochen, B. Geigl. (2015). "Circulating tumor DNA as a liquid biopsy for cancer." *Clinical Chemistry*, *61*, 112–123. doi:10.1373/clinchem.2014.222679.

[12] Li, Yuanyuan. & Trygve, O. Tollefsbol. (2011). "DNA methylation detection: Bisulfite genomic sequencing analysis." *Methods in Molecular Biology*, *791*, 11–21. doi:10.1007/978-1-61779-316-5_2.

[13] Frommer, Marianne., Louise, E. Mcdonald., Douglas, S. Millar., Christina, M. Collis., Fujiko, Watt., Geoffrey, W. Grigg., Peter, L. Molloy. & Cheryl, L. Paul. (1992). "A genomic sequencing protocol that yields a positive display of 5-methylcytosine residues in individual DNA strands." *Proceedings of the National Academy of Sciences of the USA*, *89*, 1827–31. doi:10.1073/pnas.89.5.1827.

[14] Randa, Keith., Wenjia, Qub., Thu, Hoa., Susan, J. Clark. & Peter, Molloy. (2002). "Conversion-specific detection of DNA methylation using real-time polymerase chain reaction (ConLight-MSP) to avoid false positives." *Methods*, *27*, 114–120.

[15] Xiong, Zhenggang. & Peter, W. Laird. (1997). "COBRA: a sensitive and quantitative DNA methylation assay." *Nucleic Acids Research*, *25*, 2532–4. doi: 10.1093/nar/25.12.2532.

[16] Gonzalgo, Mark L. & Peter, A. Jones. (1997). "Rapid quantitation of methylation differences at specific sites using methylationsensitive

single nucleotide primer extension (Ms-SNuPE)." *Nucleic Acids Research*, 25, 2529–31. doi: 10.1093/nar/25.12.2529.
[17] Shen, Lanlan. & Robert, A. Waterland. (2007). "Methods of DNA methylation analysis." *Current Opinion in Clinical Nutrition and Metabolic Care*, 10(5), 576-81. doi:10.1097/MCO.0 b013e3282bf6f43.
[18] Koppen, Gudrun., Amaya, Azqueta., Bertrand, Pourrut., Gunnar, Brunborg., Andrew, R. Collins. & Sabine, A. S. Langie. (2017). "The next three decades of the comet assay: a report of the 11th International Comet Assay Workshop." *Mutagenesis*, *1*, 32(3), 397-408. doi:10.1093/mutage/gex002.
[19] Collins, Andrew R. & Isabel, Gaivãoab. (2004). "DNA base excision repair as a biomarker in molecular epidemiology studies." *Molecular Aspects* of *Medicine*, 28, 307–22. doi:10.1016/j.mam.2007.05.005.
[20] Singh, Narendra P., Michael, T. McCoy., Raymond, R. Tice. & Edward, L. Schneider. (1988). "A simple technique for quantitation of low levels of DNA damage in individual cells." *Experimental Cell Research*, 175, 184–91.
[21] Epe, Bernd., Michael, Pflaum., Martin, Häring., Jutta, Hegler. & Helga, Rüdiger. (1993). "Use of repair endonucleases to characterize DNA damage induced by reactive oxygen species in cellular and cell-free systems." *Toxicology Letters*, 67, 57–72.
[22] New England Biolabs Catalog and Technical Reference, New England Biolabs, Ipswich, MA, 2005–2006, pp. 266–7.
[23] Lewies, Angélique., Etresia, Van Dyk., Johannes, F. Wentzel. & Pieter, J. Pretorius. (2014). "Using a medium-throughput comet assay to evaluate the global DNA methylation status of single cells." *Frontiers in Genetics*, 5, 215. doi:10.3389/fgene.2014.00215.
[24] Perotti, Alessio., Valeria, Rossi., Antonio, Mutti. & Annamaria, Buschini. (2015). "Comet Methy-sens and DNMTs transcriptional analysis as a combined approach in epigenotoxicology." *Biomarkers*, *20*(1), 64-70. doi: 10.3109/1354750X.2014.992813.
[25] Costa, Carla., Ana, Catarina Alves., Solange, Costa., Amadeu, M. V. M. Soares., Marta, S. Monteiro., Susana, Loureiro. & João, Paulo

Teixeira. (2015). "The use of comet assay to assess global DNA methylation in human biomonitoring studies." Frontiers in Genetics Conference Abstract: ICAW 2015 - 11th International Comet Assay Workshop. doi:10.3389/conf.fgene.2015.01.00015. https://www.frontiersin.org/10.3389/conf.fgene.2015.01.00015/event_abstract.

[26] Wentzel, Johannes F. & Pieter, J. Pretorius. (2012). "Investigating the role DNAmethylations plays indeveloping hepatocellular carcinoma associated with tyrosinemia type1 using the comet assay." *DNA Methylation-From Genomics to Technology*, edited by Tatiana Tatarinova and Owain Kerton, 211–26. Published by InTech.

[27] Wentzel, Johannes F., Chrisna, Gouwsa., Cristal, Huysamen., Etresia, van Dyk., Gerhard, Koekemoer. & Pieter, J. Pretorius. (2010). "Assessing the DNA methylation status of single cells with the comet assay." *Analytical Biochemistry,* 400(2), 190–4. doi:10.1016/j.ab.2010.02.008.

[28] Gowher, Humaira., Oliver, Leismann. & Albert, Jeltsch. (2000). "DNA of Drosophila melanogaster contains 5-methylcytosine." *EMBO Journal, 19*(24), 6918–23. doi:10.1093/emboj/19.24.6918.

[29] Zhou, Yi., Thomas, Bui., Lisa, D. Auckland. & Claire, G. Williams. (2002). "Undermethylated DNA as a source of microsatellites from a conifer genome." *Genome, 45*(1), 91–9.

[30] Townsend, Todd A., Marcus, C. Parrish., Bevin, P. Engelward. & Mugimane, G. Manjanatha. (2017). "The Development and Validation of EpiComet-Chip, a Modified High-Throughput Comet Assay for the Assessment of DNA Methylation Status." *Environmental* and *Molecular Mutagenesis,* 58(7), 508–521. doi: 10.1002/em.22101.

[31] Weingeist, David M., Jing, Ge., David, K. Wood., James, T. Mutamba., Qiuying, Huang., Elizabeth, A. Rowland., Michael, B. Yaffe., Scott, Floyd. & Bevin, P. Engelward. (2013). "Single-cell microarray enables high-throughput evaluation of DNA double-strand breaks and DNA repair inhibitors." *Cell Cycle, 12*(6), 907–15. doi:10.4161/cc.23880.

[32] Ge, Jing., Danielle, N. Chow., Jessica, L. Fessler., David, M. Weingeist., David, K. Wood. & Bevin, P. Engelward. (2015). "Micropatterned comet assay enables high throughput and sensitive DNA damage quantification." *Mutagenesis*, *30*(1), 11–9. doi:10.1093/mutage/geu063.

[33] Ge, Jing., Somsak, Prasongtanakij., David, K. Wood., David, M. Weingeist., Jessica, Fessler., Panida, Navasummrit., Mathuros, Ruchirawat. & Bevin, P. Engelward. (2014). "CometChip: a high-throughput 96-well platform for measuring DNA damage in microarrayed human cells." *Journal of Visualized Experiments*, (92), e50607. doi:10.3791/50607.

BIOGRAPHICAL SKETCH

Name: Yildiz Dincer

Affiliation: Istanbul University-Cerrahpasa, Cerrahpasa Medical Faculty, Department of Medical Biochemistry

Education: Biochemistry PhD

Business Address: Istanbul University-Cerrahpasa, Cerrahpasa Medical Faculty, Department of Medical Biochemistry, Kocamustafapasa, Fatih, Istanbul, Turkey; yldz.dincer@gmail.com

Research and Professional Experience: DNA damage and repair, antioxidant system, epigenetics

Professional Appointments: Istanbul University, Cerrahpasa Medical Faculty, Clare Hall Laboratories Cancer Research UK, Istanbul University-Cerrahpasa, Cerrahpasa Medical Faculty

Honors: Senior Professor in Istanbul University-Cerrahpasa, Cerrahpasa Medical Faculty

Publications from the Last 3 Years:

[1] Akçay, T; Yaşar, O; Kuseyri, MA; **Dincer, Y**; Aydınlı, K; Benian, A; Balcan, E; Erenel, H. Significance of serum c-erbB-2 oncoprotein,

insulin-like growth factor-1 and vascular endothelial growth factor levels in ovarian cancer. *Bratislava Med J*, 2016, 117(3), 156-160.

[2] Akkaya, Ç; Yavuzer, SS; Yavuzer, H; Erkol, G; Bozluolcay, M; **Dinçer, Y**. DNA damage, DNA susceptibility to oxidation and glutathione redox status in patients with Alzheimer's disease treated with and without memantine. *J Neurol Sci*, 2017, 378, 158–162.

[3] Himmetoğlu, Ş; Tuna, MB; Koç, EE; Ataus, S; **Dincer, Y**. Serum levels of growth factors in patients with urinary bladder cancer. *Turk J Biochem*, 2017, 42(5), 571-575.

[4] Akcay, T, Himmetoglu, S; Yasar, O; Erdem, T; Gundogdu, S; **Dincer, Y**. Evaluation of Leukocyte DNA Damage and Antioxidant Defense in Graves' Disease; Effect of Medical Treatment. *Int J Clin Pharmacol Toxicol.*, 2017, 6(3), 280-283.

[5] Himmetoglu, S; Yüksel, S; Damcı, T; İlkova, H; **Dinçer, Y**. Serum Level of Cytokeratin-18/M30 Antigen is Increased in the Cases with Impaired Glucose Tolerance. *Int Arch Endocrinol Clin Res*, 2017, Volume 3, |Issue 1, Open Access, DOI: 10.23937/2572-407X.1510011.

[6] Caliskan, Z; Mutlu, T; Guven, M; Tuncdemir, M; Niyazioğlu, M; Hacioglu, Y; **Dincer, Y**. SIRT6 Expression and Oxidative DNA Damage in Individuals with Pre-diabetes and Type 2 Diabetes Mellitus. *Gene*, 2018 Feb, 5, 642, 542-548.

[7] **Dincer, Y**; Akkaya, Ç; Alagöz, S; Pekpak, M. Assessment of Urinary Epidermal Growth Factor Level in Patients with Chronic Kidney Disease. *Urol Nephrol Open Access J.*, 2018, 6(4), 131–134.

[8] **Dincer, Y**; Himmetoglu, S; Bozcali, E. Serum level of M30 antigen in acute myocardial infarction. Focus on Sciences, 2018, 4(1).

[9] **Dinçer, Y**. Epigenetics: Mechanisms and Clinical Perspectives. **Ed. Yıldız Dincer**, Nova Science Publishers, New York, 2016, **ISBN:** 978-1-63484-519-9.

[10] **Dinçer, Y**; **Baykara, O**. Effects of Oxidative Stress on Epigenetic Mechanisms, in 'Epigenetics: Mechanisms and Clinical Perspectives', Ed. Yıldız Dincer, Nova Science Publishers, New York, 2016, p. 15-30, **ISBN:** 978-1-63484-519-9.

[11] **Dinçer, Y; Sezgin, S.** Alzheimer's Disease: An Insight from Epigenetic Perspective, in 'Epigenetics: Mechanisms and Clinical Perspectives', Ed. Yıldız Dincer, Nova Science Publishers, New York, 2016, p. 83-113, **ISBN:** 978-1-63484-519-9.

[12] **Dinçer, Y.** Alzheimer's Disease Current and Future Perspectives, **Ed. Yıldız Dincer**, OMICS Group eBooks, Foster City, California, 2016, **ISBN:** 978-1-63278-067-6.

[13] Sezgin, Z; **Dinçer, Y.** Pathogenesis of Alzheimer's Disease, in 'Alzheimer's Disease Current and Future Perspectives' **Ed. Yıldız Dincer**, OMICS Group eBooks, Foster City, California, 2016, 23-45, **ISBN:** 978-1-63278-067-6.

[14] **Dinçer, Y.** Nörobiyokimya ve Nörodejeneratif Hastalıklar, Sorularla Konu Anlatımlı Tıbbi Biyokimya. Ed. Dildar Konukoğlu, Nobel Tıp Kitabevleri, İstanbul, 2016, 627-640, **ISBN:** 978-605-335-242-6.

[15] **Dinçer, Y.** Kanser Biyokimyası ve Tümör Belirteçleri, Sorularla Konu Anlatımlı Tıbbi Biyokimya. Ed. Dildar Konukoğlu, Nobel Tıp Kitabevleri, İstanbul, 2016, 641-647, **ISBN:** 978-605-335-242-6.

[16] **Dinçer, Y; Akkaya, Ç.** *From DNA Methylation Signature in Circulating DNA to Cancer Detection and Monitoring, in 'Horizons in Cancer Research*, Volume 66', Ed. Hiroto S. Watanabe, Nova Science Publishers, New York, 2017; 117-138, **ISBN:** 978-1-53611-011-1.

[17] **Dinçer, Y.** Enzyme Kinetics and Enzymatic Analysis, in Biochemistry Laboratory Practise Book-I, İstanbul Üniversitesi Yayınları, 2017, 95-103, ISBN: 978-605-07-0640-6.

[18] **Dinçer, Y.** Research on New Generation Tumor Markers, Ed. Yıldız Dincer, Nova Science Publishers, New York, 2018, ISBN: 978-1-53614-367-6.

[19] Tuz, AA; Yüksel, S; Dinçer, Y. Commonly Used Tumor Markers and Their Limitations, in 'Research on New Generation Tumor Markers', Ed. Yıldız Dincer, Nova Science Publishers, New York, 2018, p. 1-22, ISBN: 978-1-53614-367-6.

[20] **Dinçer, Y.** Pharmacoepigenetics of memantine in dementia, in Pharmacoepigenetics Volume 10, Ed. Ramon Cacabelos, 1st edition, Elsevier (Academic Press), New York, 2019, chapter 32, Hardcover ISBN: 9780128139394 chapter 32.

In: A Closer Look at the Comet Assay
Editor: Keith H. Harmon

ISBN: 978-1-53611-028-9
© 2019 Nova Science Publishers, Inc.

Chapter 6

COMET ASSAY: A SUITABLE METHOD FOR *IN VITRO* GENOTOXICITY ASSESSMENT USING ANIMAL LYMPHOCYTES

Simona Koleničová, Viera Schwarzbacherová[*], Beáta Holečková and Martina Galdíková
Institute of Genetics, University of Veterinary Medicine and Pharmacy, Košice, Slovak republic

ABSTRACT

Neonicotinoids make up a widely-used group of insecticides which are under increased regulatory control because of their harmful impact on non-target organisms. Sensitive, simple and rapid test systems are required for pesticide risk assessment and biomonitoring studies. The *in vitro* Comet assay is now recommended as a suitable test in the technical guidance documents for the Registration, Evaluation and Authorisation of Chemicals (REACH), and is widely used in genotoxicity testing of pesticides, nanoparticles and pharmaceuticals. Comet assay under alkaline (DNA single- and double-strand break detection) or neutral (DNA double-strand break detection) conditions has been widely

---

[*] Corresponding Author's Email: vierka.kolesarova@gmail.com.

performed on cultured cells or cell lines for DNA damage assessment. DNA damage is frequently used as a biomarker connected with exposure, the process of ageing, diseases and cancer development. Here we describe and analyse the results of *in vitro* treatment of lymphocytes with insecticide using Comet assay under alkaline and neutral conditions. We tested the commercial product Calypso® 480SC and its active agent thiacloprid at concentrations of 30; 60; 120; 240 and 480 $\mu g.ml^{-1}$. The experiments were performed on isolated bovine lymphocytes, which were treated with insecticide for 2 h in two different ways: immediately after isolation and for the last 2h of 48h cultivation in cell culture media. Cattle and ruminants represent a suitable experimental model for genotoxicity assessment because of their higher exposure to pollutants than other animals through their diet.

First the impact of commercial product Calypso® 480SC was evaluated. Results of *alkaline Comet assay* showed significantly elevated DNA damage in the concentration range from 60 to 480 $\mu g.ml^{-1}$ ($p < 0.01$ and $p < 0.001$) after 2h treatment. The experiments done with pre-cultivation of lymphocytes before 2h treatment produced an increase in DNA strand breaks ($p < 0.05$) only at the highest tested concentrations (240 and 480 $\mu g.ml^{-1}$). *Neutral Comet assay* showed statistical significance only at the highest concentration tested (480 $\mu g.ml^{-1}$, $p < 0.05$) in both procedures, with and without pre-cultivation of lymphocytes. Next we analysed the results after exposure to pure active agent thiacloprid. Under *alkaline* conditions, increased levels of DNA strand breaks were detected at the highest concentration tested (480 $\mu g.ml^{-1}$; $p < 0.05$ and $p < 0.05$) in the basic and pre-cultivation procedure. *Neutral* Comet assay did not record any statistically-significant extent of DNA damage at any of the concentrations tested. Our results suggest that commercial insecticide is able to alter genetic material, and Comet assay was found to be appropriate for early detection of DNA damage.

Keywords: comet assay, DNA strand breaks, genotoxicity, neonicotinoid insecticide, viability

1. INTRODUCTION

1.1. Comet Assay

Genotoxicity is defined as the process by which an agent produces a deleterious effect on DNA and other cellular targets controlling the integrity of genetic material; this includes induction of DNA adducts, strand breaks, point mutations and structural and numerical chromosomal changes (Gollapudi and Krishna 2000). In the case of chemical substances, genotoxic agents are those that cause structural alterations in DNA, causing changes or rearrangements in the genes, thus inducing mutations (López et al. 2012). Mutations are permanent heritable changes: somatic mutations can be passed on to other cells during mitosis; mutations in gametes (germ cells) can be passed on from parent to offspring. The relationship between the accumulation of mutations in mammalian cells and the induction of neoplastic processes is generally accepted at present. For this reason, over the years increased attention has been paid to the development and validation of appropriate tests for genotoxicity assessment. These tests include *in vitro* and *in vivo* methods; among them the chromosomal aberration (CA) test, micronuclei (MN) test, sister chromatid exchanges (SCEs) detection and comet assay are very well known.

Comet assay has been used for many years as a sensitive method for assessment of DNA damage in single cells; the test is also known as Single Cell Gel Electrophoresis, SCGE. This assay allows the measurement of double-strand and single-strand breaks in DNA both *in vivo* and *in vitro*, which may be due to DNA interaction with genotoxic substances. For this reason the test is used to detect DNA damage as an indicator of exposure to genotoxic agents (Collins et al. 2014; Gharsalli 2016). It is a highly versatile assay; the large number of studies on various types of genotoxic agents makes it an ideal screening test (Moller 2018). The principle of the method depends on the relaxation of supercoiled DNA in agarose-embedded nucleoids, with residual bodies remaining after lysis of cells with detergent and high salt, which allows the DNA to be drawn out towards the anode under electrophoresis, forming comet-like images as seen under fluorescence microscopy (Azqueta and Collins 2013). According to the conditions in which the comet test is performed, it is divided into alkaline and neutral. The alkaline test is the most widely used for measuring DNA damage in eukaryotic cells (Neri et al. 2015). Langie et al. (2015) noted that in eukaryotic cells the assay detects a biologically useful range of strand breaks (SB) and alkali-labile sites (from several hundred to several thousand breaks per cell). One explanation for the success of this method is that electrophoresis of lysed cells under alkaline conditions allows partial disruption of DNA secondary structure and removes DNA tertiary and quaternary structure. Alkaline conditions also degrade RNA and reveal more DNA lesions, including single-strand breaks, double-strand breaks and alkali-labile sites, so they are more sensitive than neutral conditions which reveal only double-strand breaks (Singh 2016).

The physical principles of DNA track formation during electrophoretic migration of DNA have been unclear for a long time. Recently, Afanasieva et al. (2018) used an original approach based on kinetic measurements of the comet formation. They found that in alkaline conditions, linear DNA fragments make an essential contribution to the tail formation; in the neutral comet assay, the tail is formed by extended DNA loops which are about the same as chromatin loops in the cell nuclei.

Besides measuring DNA breaks, the comet test has also been adapted to detect alkaline labile sites, oxidative DNA damage, DNA cross-links, DNA adducts, apoptosis and necrosis (Bajpayee et al. 2013); however, use of comet assay in the study of apoptosis is a subject of discussion. Collins et al. (2008) explain that apoptosis is not an immediate consequence of a severe assault on the cell, but takes time to develop. Moreover, breaks induced by agents are usually repaired, in contrast to breaks occurring under apoptosis, which are irreversible. Thus breaks caused by apoptosis-inducing treatment and revealed using comet assay represent the earliest stages of apoptosis. In contrast, hedgehog comets (comets with a small nucleoid head and large broad tail) cannot be used as a specific indicator of apoptosis (Collins et al. 2008; Lorenzo et al. 2013; Azqueta and Collins 2013; Koppen et al. 2017). According to Kuchařová et al. (2018), one of the essential applications of comet assay is the detection of oxidative damage to DNA, as a significant contributor to human diseases.

The source of cells for comet assay may be blood (human, animal) containing lymphocytes. Peripheral blood lymphocytes (PBL) can be easily collected in a relatively non-invasive way for determination of the degree of DNA damage using comet testing (Kuchařová et al. 2019). The advantages and disadvantages of using lymphocytes have been reported in the review by Collins et al. (2008): they are easily obtained in large numbers; they are diploid and are almost all in the same phase of the cell cycle (G_0); as they circulate through the whole body, they can be seen as reflecting the overall state of the organism. The main disadvantages of using lymphocytes include: they achieve very limited survival *in vitro*, they display individual phenotypic variation, and they do not reflect the level of damage in a particular organ with specific metabolism. As emphasized by Bajpayee et al. (2013), the critical point is to obtain viable cells to avoid ambiguity in the interpretation of the range of DNA damage. Usually, 100 random cells from each slide per sample are scored under a fluorescent microscope linked to a charge-coupled device (CCD) camera to acquire images using the software. The most commonly-used comet parameters recorded are olive tail moment (OTM, arbitrary units), tail DNA (%) and tail length (TL, distance of migration of the DNA from the nucleus; μm).

1.2. Neonicotinoid Pesticides

Neonicotinoids are one of the most important classes of insecticides chemically related to nicotine. They include imidacloprid, acetamiprid, thiacloprid, dinotefuran, nitenpyram, thiamethoxam, and clothianidin. They are the fastest-growing class of insecticides worldwide and are widely used for plant protection in agriculture in more than 120 countries (Jeschke et al. 2011). They have also been commonly used in veterinary medicine as an effective flea treatment for dogs and cats (Tomizawa and Casida 2005). Their popularity is due to the unique biological and chemical properties which they possess, such as broad-spectrum insecticidal activity, low application rates, excellent uptake and translocation in plants, a new mode of action and favourable safety profile (Maienfisch et al. 2001). Neonicotinoids are systemic; they have long persistence and are highly water-soluble. Unfortunately, however, all of these properties lead to the possibility of environmental contamination and adverse effects on non-target organisms (Morrissey et al. 2015; Wood and Goulson 2017; Berheim et al. 2019).

These insecticides act as neurotoxins on the insect nervous system by interacting with the insect acetylcholine receptor (AChR) (Blacquiére et al. 2012), which leads to the accumulation of acetylcholine, resulting in paralysis and death of the insect (Iwasa et al. 2004). The problem with neonicotinoids lies in their interaction with non-target insects, for instance honeybees. Exposure to lower concentrations of neonicotinoids leads to sublethal effects such as impaired learning and memory, impaired immune defence of honeybees, or negative impact on the reproductive anatomy and physiology of honeybee queens (Lundin et al. 2015; Williams et al. 2015; Brandt et al. 2016). Considering the confirmed harmful effects of neonicotinoid pesticides on honeybees in 2018, the EU banned the three main neonicotinoids, clothianidin, imidacloprid and thiamethoxam, for all outdoor uses.

Several studies have shown the deleterious effects of neonicotinoid insecticides on human and animal health as well. Sekeroglu et al. (2013) investigated the genotoxic and cytotoxic potential of commercial formulations of thiacloprid and deltamethrin alone or in combined treatment applied to rat bone marrow cells. They found significantly increased frequencies in MNi and CA. In another study, Galdíková et al. (2015) confirmed the ability of a commercial formulation of thiacloprid (Calypso® 480SC) to induce genotoxic effects in cultured bovine peripheral lymphocytes *in vitro*. After 24 h treatment with thiacloprid formulation statistically significant increases in CAs and DNA damage were found at doses ranging from 120 to 480 $\mu g.ml^{-1}$. After 48 h treatment, elevation in SCE induction was observed at concentrations ranging from 120 to 480 $\mu g.ml^{-1}$.

The genotoxic potential of imidacloprid has been shown *in vivo* in bone marrow chromosome aberration assay and micronucleus testing after treatment with imidacloprid with concentrations of 50 and 100 $mg.kg^{-1}$ in rats (Karabay and Oguz 2005). Similarly, Demsia et al. (2017) observed statistically significant elevation in MNi formation at the highest dose in *in vivo* micronucleus assay with rat bone-marrow polychromatic erythrocytes. However, they did not confirm any positive effect of imidacloprid on human lymphocytes *in vitro* in MN and SCE assay within the same study. Costa et al. (2009) investigated the induction of DNA damage using comet assay and testing for micronuclei formation in human lymphocytes. They found a significant increase in both analyses only at the highest tested concentration of imidacloprid (20 µM). Similar results have been observed by other authors, such as Feng et al. (2005), in human lymphocytes *in vitro*. Lower concentrations of imidacloprid did not produce any positive effect in SCE and MN induction. Significant elevation in DNA damage assessed by means of comet assay was observed in all treated groups. Al-Sarar et al. (2015) studied the effects of imidacloprid on Chinese hamster ovary cells. They observed significant elevation of frequency of micronuclei, however no increase in chromosome or DNA damage. Conversely, no significant differences in MN frequencies were found in human peripheral lymphocytes after exposure to imidacloprid (Stivaktakis et al. 2010).

Another member of the neonicotinoids, clothianidin, was studied in human peripheral lymphocytes with and without a metabolic activation system. Researchers observed genotoxic and cytotoxic effects of clothianidin using CA and MN assay as well as mitotic and nuclear division indices (Sekeroglu et al. 2018). In another *in vivo* study, the genotoxic potential of a commercial formulation of clothianidin (Poncho®) was demonstrated in CD1 male mice using micronucleus and comet assay.

The genotoxic and cytotoxic potential of acetamiprid was confirmed in the study by Cavas et al. (2012), where the effects of the insecticide were evaluated on human intestinal CaCo-2 cells using micronucleus, comet and γH2AX foci formation assays. Similarly, significant induction of micronuclei and DNA damage measured using alkaline comet and γH2AX foci formation assays was found in a human lung fibroblast cell line (Cavas et al. 2014).

In this study we have used alkaline and neutral comet assay to assess the potential genotoxic effects of thiacloprid and its commercial formulation Calypso® 480 SC on bovine lymphocytes after exposure for 2 h (with and without pre-cultivation).

2. METHODS

2.1. Materials

2.1.1. Materials for Cell Cultures
- RPMI 1640 with L-glutamine supplement and 25 mM HEPES (GE Healthcare Hyclone Lab, Utah, USA), phytohemagglutinin (PHA-L, 20 µg.ml^{-1}, PAN Biotech, Germany), Histopaque®-1077 (Sigma-Aldrich, St. Louis, MO, USA), phosphate-buffered saline (Oxoid, Basingstoke, Hampshire, England).
- Thiacloprid (C$_{10}$H$_9$ClN$_4$S, CAS Registry 111988-49-9, > 99% purity, Sigma-Aldrich, St. Louis, MO, USA) dissolved in water and applied to culture media at concentrations of 30, 60, 120, 240 and 480 µg.ml^{-1}.
- Thiacloprid-based insecticide, trade name Calypso® 480SC, with 480 g.l^{-1} of active agent **[3-(6-Chloro-3-pyridinylmethyl)-2-thiazolidinylidene] cyanamide** (Sigma-Aldrich, St. Louis, MO, USA).
- Other chemicals were purchased from Sigma-Aldrich (St. Louis, MO, USA).
  1. Blood sampling: two healthy bull donors (Slovak spotted cattle, 5-6 months old), *v. jugularis*, anticoagulant heparin-coated syringe (12 ml).
  2. RPMI 1640 medium supplemented with L-glutamine, 25 mM HEPES and sterile water, 500 ml packet. Stored at 2 – 8°C.
  3. Fetal bovine serum: ready-to-use solution.
  4. Antibiotic and antimycotic mix solution (100 U.ml^{-1} penicillin, 0.1 mg.ml^{-1} streptomycin and 0.25 µg.ml^{-1} amphotericin).
  5. PHA-L Phytohemagglutinin, 1.2 mg protein/flask, lyophilized, 6x5 ml. Stored at 2 – 8 °C.

6. Lymphocyte separating medium: Histopaque®-1077 for isolations from whole blood. Ready-to-use solutions. Stored at 4 – 8°C.
7. Phosphate-buffered saline (Dulbecco A), 100 tablets, add 1 to 100 ml distilled water (ddH$_2$O), pH = 7.3 ± 0.2. Stored at 25°C.
8. Trypan blue 0.4% ready-to-use solution.
9. Thiacloprid (active substance), CALYPSO® (commercial preparation). Stock solutions: thiacloprid – 480 µg.ml^{-1} – 48 mg dissolved in 1 ml ddH$_2$O. CALYPSO® – 100 µl dissolved in 900 µl ddH$_2$O. Concentrations of neonicotinoids used: 30, 60, 120, 240 and 480 µg.ml^{-1}.
10. Hydrogen peroxide, positive control. Stock solutions: 0.1 M H$_2$O$_2$ – 50 µl 10 M H$_2$O$_2$ dissolved in 4.95 ml PBS. For use 1x: 300 µM H$_2$O$_2$ – 300 µl 1 M H$_2$O$_2$ dissolved in 700 µl ddH$_2$O.
11. Centrifuge, serological pipettes, glass tubes (10 ml), plastic tubes (15 ml), beakers, Burker chamber.

2.1.2. Comet Assay Components
- Low melting-point agarose (LMPA, SERVA Electrophoresis GmbH, Heildelberg, Carl-Benz-Str.7, Germany).
- Other chemicals were purchased from Sigma-Aldrich (St. Louis, MO, USA).

2.1.2.1. Alkaline Comet Assay
1. CometSlides™ 2 Well
   Slides with two sample surfaces specially treated to promote agarose adherence and a hydrophobic barrier, greatly simplify the comet assay by providing two specially-treated sample surfaces to allow easy application of cells directly to each slide. One hundred slides. Stored at room temperature.
2. Microscope cover glass – 24x40 mm.
3. 0.75% low melting-point agarose (LMPA) in PBS, 25g. Storage at 15 – 30°C.

4. Lysing solution

   2.5 M NaCl, 0.1 M, disodium ethylene-diaminetetraacetic acid (EDTA disodium salt), 10 mM Trizma base. 146.1 g of NaCl, 37.2 g of EDTA and 1.2 g of Trizma added to 700 ml of ddH$_2$O, and mixture stirring initiated. pH adjusted to 10 by first adding crystalline NaOH granules to pH = 8.0 and finally adding 1 M NaOH. Solution stored at 4°C. Final lysis solution.

5. Lysis solution – pH 10 and 1% Triton X-100. Mixed for 30 minutes at 4°C before use.

6. Electrophoresis buffer – 0.3 M NaOH, 1 mM Na$_2$EDTA in ddH$_2$O. Stock solutions of 5 M NaOH prepared: 100 g dissolved in 500 ml of ddH$_2$O and 200 mM EDTA: 37.22 g dissolved in 200 ml of water.

   For use 1x working buffer: 300 mM NaOH/1 mM EDTA from the stock solution. 60 ml of 5 M NaOH and 5 ml of 200 mM EDTA added to stock solution and made up to 1 L with chilled water.

7. Neutralization buffer

   0.4 M Tris base prepared in ddH$_2$O. pH adjusted to 7.5 with 10 M HCl.

   1 M Tris prepared by dissolving crystalline Tris (MW = 121.14 g.mol^{-1}) in ddH$_2$O. pH of the solution adjusted to 7.5 with concentrated HCl; solution supplemented with water to the desired amount of 1 L and stored at room temperature. Solution prepared 30 minutes before use.

8. Staining solution

   5 µg.ml^{-1} ethidium bromide. Stock solution (1:1000 in ddH$_2$O) can be stored at 4°C.

9. Slide boxes, water bath, pipette, laboratory glassware, horizontal electrophoresis tank

10. Quantification

    Nikon ECLIPSE Ni-U epifluorescence microscope.

    Charge-coupled device (CCD) camera

    Computer Workstation

CASP image analysis software was available as a freeware version (http://casplab.com/).

2.1.2.2. Neutral Comet Assay
1. Lysing solution: 2.5 M NaCl, 0.1 M disodium ethylenediaminetetraacetic acid (EDTA disodium salt, 10 mM Tris-HCl, pH = 9.5), 1 N-lauroylsarcosine sodium salt, 1% TritonX-100.
2. Neutral electrophoresis buffer: 10x TBE buffer (tris-borate-EDTA), diluted to working concentration with distilled water. Stored at room temperature.

2.2. Methodology

2.2.1. Lymphocyte Separation for in Vitro Comet Assay
All steps should be performed under aseptic conditions.

2.2.1.1. Blood Sampling
Our blood collection from two healthy bull donors (5 – 6 months old, Slovak spotted cattle) was performed 35 – 40 minutes before lymphocyte isolation from *v. jugularis* into heparinized tubes at room temperature.

Blood collection is carried out in a stable, so it must be sterile to avoid contamination of the samples. The collection point is cut and disinfected with alcohol. The donor animals must be clinically healthy to avoid distortion of results. After taking the blood in the heparinised syringe, it is good to leave it 30 – 40 min at room temperature to settle the erythrocytes.

2.2.1.2. Culture Medium
The culture medium is prepared before further processing of the collected blood in plastic tubes (15 ml).

(a) Medium for non-proliferating lymphocytes (without cultivation, only 2 hours exposure): 4 ml RPMI 1640 medium + 1 ml bovine foetal serum + 40 µl antibiotic mixture

(b) Medium for proliferating lymphocytes (with 48 hours cultivation and exposure for the last 2 hours of cultivation): 4 ml RPMI 1640 medium + 1 ml bovine foetal serum + 40 µl antibiotic mixture + 100 µl phytohaemagglutinin
The prepared media are kept in a thermostat at 37°C.

2.2.1.3. Isolation of Lymphocytes

- Dilute whole blood 1:1 with PBS, mix blood and PBS well by inverting the sealed tubes or gently mixing on a shaker. Add Histopaque to the prepared glass tubes and slowly layer blood onto Histopaque on the tube wall. After spinning, the resulting ring from the lymphocytes is carefully aspirated so as not to aspirate the Histopaque and erythrocytes. The aspirated cells are washed with PBS. After the supernatant is collected, the resulting cell pellet is resuspended in PBS by gentle homogenization using the tip of a Pasteur pipette.
- After spinning, the supernatant is aspirated again and the resulting pellet is resuspended in PBS (Panda and Ravindran 2013).
- The cells are counted visually using the Burker chamber. Approximately 5×10^5 cells (no more) are required for negative control and used concentrations of test neonicotinoids (thiacloprid and Calypso®). A maximum of 50 µl of the prepared cells is added to the culture medium to obtain the required amount.
- The heated culture medium in the thermostat is supplemented with the same volume of lymphocytes isolated in all tubes. Lymphocytes are exposed to the test insecticides for two hours, concurrently with their addition to the medium (non-proliferating lymphocytes); lymphocytes with cultivation are stored for 48 hours in the thermostat and exposed for the last 2 hours (proliferating lymphocytes). Thiacloprid and Calypso 480® are pre-diluted, as well as the individual concentrations which are added to the cells in the media at a volume of 50 µl.

- After cultivation is completed, the tubes are centrifuged, the supernatant is aspirated and cells are washed with PBS. Cells are resuspended and centrifuged. The supernatant is aspirated, and the pellet is resuspended in a volume of approx. 1 ml.

2.2.1.4. Cytotoxicity
- Trypan blue dye exclusion testing is carried out to assess cell viability. 20 µl trypan blue dye is added to 20 µl of the cell suspension and mixed. After 5 min the numbers of cells are counted: dead cells are recognized as blue and live cells as shiny in the Burker chamber. The percentage viability and number of cells left after treatment are calculated. The necessary amount of cells in PBS is transferred to the prepared Eppendorf tubes.

2.2.2. Comet Assay

2.2.2.1. Alkaline Comet Assay

2.2.2.1.1. Slide Preparation and Agarose Embedding
In our study the procedure of Singh et al. (1988) was applied, with modifications by Slameňová et al. (1997) and Gábelová et al. (1997).

- Prepare specially modified CometSlides™ and cover glasses for each concentration, on a black background. Slides are coated with special agarose to replace 1% NMPA.
- Melt 1% and 0.5% LMPA and maintain at 37°C in a water bath.
- Mix the cell suspension containing the cells of interest in PBS with 0.75% LMPA. 500 µl of LMPA is added to the cells.
- 85 µl of the mixture is added to each prepared Comet slide™. Place coverslips (24 x 40 mm) and put slides on a sealable tray coated with filter paper soaked with water to the refrigerator.

Caution: Avoid creating any air bubbles during mixing as this can result in artificial DNA damage.

2.2.2.1.2. Lysing

Once the agarose is set, carefully remove the coverslips and place the slides vertically in a cuvette, freshly prepared with alkaline lysis solution. Lysing takes place in the refrigerator at 4°C for 1 h.

2.2.2.1.3. Unwinding

The lysis solution is discarded, and the slides are gently removed from the cuvette. The slides are placed next to each other in an electrophoretic tank. The prepared slides are covered with fresh, cold electrophoretic buffer (pH > 13) until it completely covers the slides (avoid bubbles over the agarose). Place the tank in a closed refrigerator set at 4°C to prevent melting of the agarose layer during electrophoresis.

2.2.2.1.4. Electrophoresis

Electrophoresis of DNA is performed at a constant voltage of 25V (0.75 V.cm^{-1}) with the current reaching 300 mA for 30 min. When starting electrophoresis, individual values need to be checked. In the case of fluctuating current or voltage, remove or add alkaline electrophoretic buffer using a 10 - 50 ml syringe.

2.2.2.1.5. Neutralization

Turn off the power source. Carefully remove the slides from the tank so that the gel is not damaged and put the slides in a container filled with paper towels to dry. Drop-wise coat the slides with neutralization buffer (pH = 7.4) in the cuvette. This process is repeated two more times for 10 min or three times for 5 min.

2.2.2.1.6. Staining

After removing the solution, the slides are transferred to the paper towel container again. Stain the electrophoresed DNA with 15 µl of ethidium bromide solution (both sides of each slide) and cover with a coverslip (Horváthová et al. 2006).

2.2.2.1.7. Visualization of DNA Damage

Wipe the slide and coverslip carefully before viewing under the microscope. Slides are observed at 20x and 40x magnification under a fluorescent microscope. 50 − 100 random cells are scored on each slide. The microscope used to assess the quantitative and qualitative cell damage is linked to a charge-coupled device (CCD) camera. The most frequently observed comet parameters are % DNA in the tail, olive tail moment (OTM, arbitrary units) and tail length. Finally, the % DNA in the tail parameter can be analyzed using the free version of CASPLab software, where each cell is evaluated separately (Końca et al. 2003).

2.2.2.2. Neutral Comet Assay

There are several changes in the neutral variant compared to the alkaline comet analysis procedure.

1. Lysis solution composition (2.5 M NaCl, 0.1 M disodium ethylenediaminetetraacetic acid (EDTA disodium salt), 10 mM Tris-HCl, pH = 9.5, 1% N-lauroylsarcosine sodium salt, 1% Triton X-100). The cells were lysed with lysis solution at 4°C for 1 h.
2. Electrophoresis was performed at 20V for 40 min in TBE buffer (Gyori et al. 2014).

2.2.3. Statistics

All microscope slides were coded before their analysis. The results are presented as mean ±SD, where n = 3. Simple analysis of variance (ANOVA) and Dunnet's *a posteriori* test were used to assess differences between control and insecticide-treated cells. Next, ANOVA and Tukey´s *a posteriori* test were used to assess differences between cultivation periods (2 and 48 h).

3. RESULTS

The results of our analysis of DNA damage using alkaline comet assay in non-proliferating and proliferating lymphocytes from bovine peripheral blood after exposure to pure neonicotinoid insecticide thiacloprid at concentrations of 30, 60, 120, 240 and 480 µg.ml^{-1} are summarized in Figure 1. DNA damage after exposure to the commercial thiacloprid preparation Calypso® 480SC at concentrations of 30, 60, 120, 240 and 480 µg.ml^{-1} is summarized in Figure 2. The percentage viability of non-proliferating and proliferating lymphocytes from bovine peripheral blood following exposure to thiacloprid and Calypso® is shown in Figures 5 and 6 respectively.

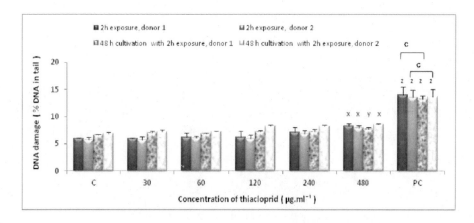

Figure 1. Percentages of DNA in tail estimated by means of alkaline comet assay in bovine peripheral blood lymphocytes (non-proliferating and proliferating) treated with thiacloprid for 2 h.

The proportion of double-stranded breaks detected by means of neutral comet analysis in non-proliferating and proliferating lymphocytes after exposure to thiacloprid at the same concentrations is shown in Figure 3, and after exposure to Calypso® also at the same concentrations in Figure 4. Viability of non-proliferating and proliferating lymphocytes from bovine peripheral blood following exposure to thiacloprid and Calypso® is shown in Figures 7 and 8 respectively.

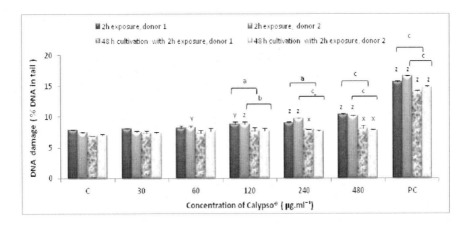

Figure 2. Percentages of DNA in tail estimated by means of alkaline comet assay in bovine peripheral blood lymphocytes (non-proliferating and proliferating) treated with Calypso® for 2 h.

Alkaline comet analysis after 2h exposure of non-proliferating lymphocytes to thiacloprid showed an increase in DNA damage with statistical significance at the highest concentration in donor 1 (480 µg.ml^{-1}; $p < 0.05$; Figure 1) and donor 2 (480µg.ml^{-1}; $p < 0.05$; Figure 1). After 48h cultivation and exposure to thiacloprid for the last 2 h, DNA damage was observed in proliferating lymphocytes with statistical significance in donor 1 (480 µg.ml^{-1}; $p < 0.01$; Figure 1) and donor 2 (480µg.ml^{-1}; $p < 0.05$; Figure 1). The viability of non-proliferating and proliferating lymphocytes in both donors exposed to thiacloprid was greater than 86.67% (Figure 5).

DNA damage to non-proliferating lymphocytes after exposure to the commercial preparation Calypso® with thiacloprid as active ingredient was observed in donor 1 at 120, 240 and 480 µg.ml^{-1} (for 120 µg.ml^{-1} $p < 0.01$; for 240 and 480 µg.ml^{-1} $p < 0.001$; Figure 2) and donor 2 at 60, 120, 240 and 480 µg.ml^{-1} (for 60 µg.ml^{-1} $p < 0.01$; for 120, 240 and 480 µg.ml^{-1} $p < 0.001$; Figure 2). Proliferating lymphocytes exposed to this formulation displayed DNA damage in donor 1 at the highest tested concentrations of 240 and 480 µg.ml^{-1} ($p < 0.05$; Figure 2) and in donor 2 at 480 µg.ml^{-1} ($p < 0.05$; Figure 2). The viability of non-proliferating and proliferating lymphocytes after exposure to Calypso® in both donors was higher than 86.7% (Figure 6).

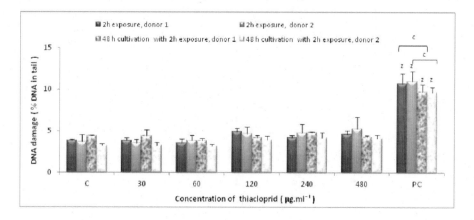

Figure 3. Percentages of DNA in tail estimated by means of neutral comet assay in bovine peripheral blood lymphocytes (non-proliferating and proliferating) treated with thiacloprid for 2 h.

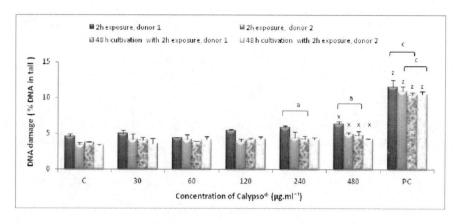

Figure 4. Percentages of DNA in tail estimated by means of neutral comet assay in bovine peripheral blood lymphocytes (non-proliferating and proliferating) treated with Calypso® for 2 h.

Statistically-significant increases in DNA damage with double-stranded breaks in proliferating and non-proliferating lymphocytes were detected by means of neutral comet assay after exposure to thiacloprid (Figure 3) and Calypso® (Figure 4) at the same concentrations as for the alkaline comet assay. Lymphocyte viability is shown in Figure 7 and 8.

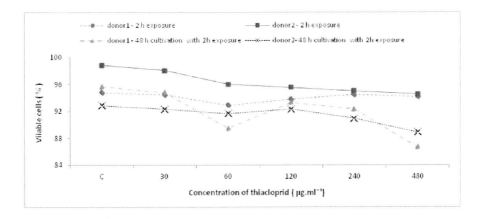

Figure 5. Viability of bovine peripheral blood lymphocytes treated with insecticide thiacloprid *in vitro* for 2 h and used in alkaline comet assay.

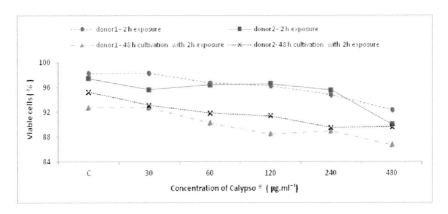

Figure 6. Viability of bovine peripheral blood lymphocytes treated with insecticide Calypso® *in vitro* for 2 h and used in alkaline comet assay.

DNA damage results detected by neutral comet analysis after exposure of non-proliferating and proliferating lymphocytes to thiacloprid did not indicate any statistically-significant DNA damage in donor 1 or donor 2 (Figure 3). Lymphocyte viability found in both donors after thiacloprid exposure was higher than 83.3% (Figure 7).

After exposure of non-proliferating and proliferating lymphocytes to Calypso®, we observed statistically-significant DNA damage at the highest concentration in both donors (480 µg.ml^{-1}; $p < 0.05$; Figure 4). Viability in donors 1 and 2 with culture and also without culture after exposure of Calypso® was higher than 83.3% (Figure 8).

Comparing the effects of individual concentrations of thiacloprid on bovine peripheral blood lymphocytes after 2 h exposure and 48 h cultivation with exposure for the last 2 h, we found that DNA damage was not statistically significant in either donor (Figure 1), with single-strand breaks detected only in alkaline comet analysis. Neutral comet analysis also showed no statistically-significant DNA damage (Figure 3).

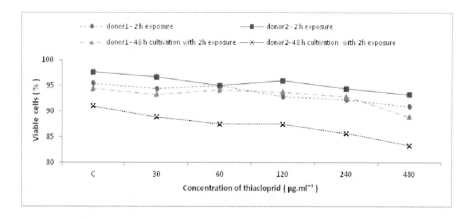

Figure 7. Viability of bovine peripheral blood lymphocytes treated with insecticide thiacloprid *in vitro* for 2 h and used in neutral comet assay.

In the alkaline comet assay variant, there was statistically-significant DNA damage after exposure to the commercial preparation, found after comparison of identical concentrations in proliferating and non-proliferating lymphocytes in donor 1 at 120, 240 and 480 µg.ml^{-1} (for 120, 240 µg.ml^{-1} $p < 0.05$; for 480 µg.ml^{-1} $p < 0.001$; Figure 2) and donor 2 at 120, 240 and 480 µg.ml^{-1} (for 120 µg.ml^{-1} $p < 0.01$; for 240 and 480 µg.ml^{-1} $p < 0.001$; Figure 2). Calypso® exposure followed by neutral comet analysis revealed statistically-significant damage only in donor 1 at concentrations of 240 and 480 µg.ml^{-1} ($p < 0.05$; Figure 4).

In the figures, "C" is control, "PC" is positive control, and *p* is supplemented with *a, b, c* to show the differences between cultivation times and *x, y, z* to point out the differences between control and individual concentrations:

* *(a < 0.05)*, ** *(b < 0.01)*, *** *(c < 0.001)* — differences between cultivation times (2 and 48h), ANOVA and Tukey´s test;

* *(x < 0.05)*, ** *(y < 0.01)*, *** *(z < 0.001)* — differences between control and individual concentrations, ANOVA and Dunnet's test.

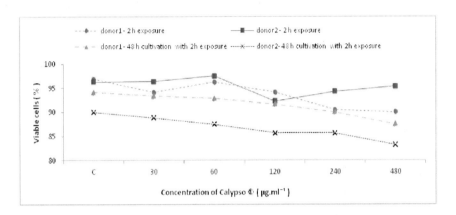

Figure 8. Viability of bovine peripheral blood lymphocytes treated with insecticide Calypso® *in vitro* for 2 h and used in neutral comet assay.

4. DISCUSSION AND CONCLUSION

Pesticides, a wide-ranging group of chemicals acting in different ways, have increased yields of agricultural products and enhanced the control of transmittable plant diseases. On the other hand, pesticides have become known for their harmful effects on non-target organisms, persistence in the environment and its contamination. Long-term exposure to pesticides is accompanied with disruption of functions of different organs, and there is evidence of links between pesticide exposure and human chronic diseases such as cancer, Alzheimer, Parkinson and diabetes mellitus (De Souza et al. 2011; Mostafalou and Abdollahi 2012). Nowadays, chronic exposure to low doses of pesticides is considered to be one of the main factors potentiating the development of cancer. Pesticides can affect genetic material in two ways: either they cause structural or functional damage to DNA and chromosomes, or the impact on cells and cell organelles disrupts their gene expression (George and Shukla 2011).

Comet assay is a widely-used method in human and animal studies for detecting the potential genotoxicity of pesticides, drugs, chemicals, nanoparticles and food additives at the level of a single cell. Based on the experimental conditions, this assay is able to detect DNA single-strand breaks (alkaline version), DNA double-strand breaks (neutral and alkaline version), DNA cross links and apoptosis (Singh 2000). While *in vivo* mammalian alkaline comet assay is recommended for genotoxic hazard assessment of chemicals by the European Food Safety Authority and has its own OECD guideline, for *in vitro* comet assay validation studies are ongoing, and this assay is recommended for use in the technical guidance documents for the Registration, Evaluation and Authorisation of Chemicals (REACH). Moreover, there is significant potential in *in vitro* comet assay analysis. It could be a future tool in the genotoxicity field in the sense of detecting a wider range of damage and creating new rapid automated scoring methods and systems (Koppen et al. 2017).

In our experiments, we used bovine lymphocytes isolated from whole bovine blood and treated them with the tested insecticides. Lymphocytes represent a suitable model for genotoxicity testing; it is easy to obtain them using a relatively non-invasive method, they have a stable genome, and they work well in comet assay (Zeljezic et al. 2016) and other genotoxicity assays (e.g., chromosomal aberration test, micronucleus assay). Cattle are chosen as a source of lymphocytes because of the high probability/risk of their being exposed to pesticides via their feed during grazing (Willet et al. 1993). Then some of the pesticides are deposited in their body fat, or are excreted through the milk and may have a harmful impact on calves and ultimately on the final consumers of animal products, people (Jandacek and Tso 2001; Kalantzi et al. 2001). Moreover, genotoxicity testing of pesticides and mycotoxins (Lioi et al. 1998; Lioi et al. 2004; Galdíková et al. 2015; Schwarzbacherová et al. 2017) has been widely performed on bovine lymphocytes.

The possible genotoxic effects of thiacloprid-based insecticide formulation Calypso® 480SC and its active agent (pure thiacloprid) were assessed in bovine lymphocytes using alkaline and neutral types of comet assay. Neonicotinoids act-selectively as agonists of nicotinic acetylcholine receptors in the central nervous system of insects (Han et al. 2018). Additionally, treatment was performed on proliferative and non-proliferative lymphocytes to evaluate whether the status of cells has an impact on the DNA damage level.

Our results show that Calypso® is able to induce significant levels of DNA damage. Firstly the alkaline version of comet assay was carried out. After thiacloprid treatment, significant levels of DNA damage were observed in both proliferative and non-proliferative lymphocyte cells, but only at the highest concentration tested (480 µg.ml^{-1}; $p < 0.05$ and $p < 0.01$). On the other hand, Calypso® exposure induced increased levels of DNA damage at all concentrations ranging from 60 to 480 µg.ml^{-1} ($p < 0.01$ and $p < 0.001$) in non-proliferative cells and at concentrations of 240 and 480 µg.ml^{-1} ($p < 0.05$) in proliferative cells. Neutral comet assay did not reveal any significant changes after pure thiacloprid treatment. Only the highest concentration of Calypso® (480 µg.ml^{-1}; $p < 0.05$) caused significant DNA damage according to this type of assay. Finally, we compared the results of non-proliferative and proliferative lymphocytes with each other to find any significant differences. Thiacloprid treatment did not cause any significant difference between proliferative and non-proliferative cells ($p > 0.05$). The major difference was seen after Calypso® exposure ($p < 0.05$; $p < 0.01$ and $p < 0.001$) from the concentration of 120 µg.ml^{-1} upwards under alkaline conditions. Comparison of results obtained from neutral comet assay showed a significant difference only in donor no. 1 (240 and 480 µg.ml^{-1}; $p < 0.05$).

Our data are in agreement with the previously-published results from bovine lymphocytes after Calypso® treatment (Galdíková et al. 2015). The insecticide was tested at concentrations ranging from 30 to 480 µg.ml^{-1}. The lymphocytes were cultured for 24 h and treated with pesticide for the last 2 h of incubation time. In comparison with our analysis, these researchers classified comets only visually into the four categories. They found increased levels of DNA damage in a dose-dependent manner from the concentration of 120 µg.ml^{-1} upwards. Similarly, significantly increased percentages of comets and tail lengths were observed after Calypso® treatment in human peripheral lymphocytes (Calderon-Segura et al. 2012). These researchers used a short treatment (2 h) with Calypso® concentrations ranging from 0.6 to 1.4 × 10^{-1} M without pre-cultivation. Changes were detected at all concentrations tested. Furthermore, the effects of pure thiacloprid were examined on the earthworm *Eisenia fetida* (Feng et al. 2015). The worms were exposed to 1 and 3 mg.kg^{-1} thiacloprid contained in dry soil for different exposure times (from 7 to 56 days). Coelomocytes were also used for comet assay experiments. The researchers detected significant DNA damage (olive tail moment), which increased with lengthening exposure time and then decreased with the recovery time. Moreover, changes in enzyme levels were assessed after exposure to thiacloprid. Activities of gluthatione-S-transferase (GST), carboxylesterase (CarE), superoxide dismutase (SOD), catalase (CAT) and peroxidase (POD) were significantly inhibited for one or more treatment times. Activity of these enzymes is often used as a marker of elevated reactive oxygen species (ROS) production (Bernard et al. 2014). Moreover, ROS production is often considered as the primary source of DNA damage after pesticide exposure (Schwarzbacherová et al. 2017; Fu et al. 2019).

It has been found that several other neonicotinoids are also able to induce DNA damage. One of them, acetamiprid, was tested on *Eisenia fetida* (Li et al. 2018). Significant changes in olive tail moment in a dose-dependent manner (0.05 – 0.50 mg.kg^{-1}) were detected. Similarly to the thiacloprid data, the researchers observed increased ROS levels and changes in antioxidant enzymes CAT and GST. Others found concentration-dependent increases in DNA damage after acetamiprid treatment (25 - 300 µM) in CaCo-2 cells (Cavas et al. 2012). Further, the formation of micronuclei and γH2AX foci were assessed. Micronuclei originate from aneugenic or clastogenic events, and γH2AX foci measure DNA double-strand breaks. It was shown that acetamiprid caused dose-dependent ($p < 0.05$ and $p < 0.01$) induction of micronuclei and γH2AX foci from the concentrations of 50 and 75 µM respectively. Next, an imidacloprid-based formulation was tested on HepG2 cells to assess its potential genotoxicity (Bianchi et al. 2015). The researchers detected elevated levels of micronuclei (of clastogenic origin) and significant DNA damage (visual scoring of comet tails) after 24 h exposure. The effects of pure imidacloprid were examined on HepG2 cells and human peripheral blood lymphocytes (Zeljezic et al. 2016). After 4 h exposure they found an increased level of DNA presence in the tail at the highest concentration tested ($p < 0.05$; 0.314 µg.ml^{-1}). Prolonged exposure time (24 h) resulted in significantly higher tail intensity ($p < 0.05$) over the concentration range. Moreover, it was shown that this insecticide was able to induce the formation of micronuclei, nucleoplasmic bridges and nuclear buds in a significant manner in both cell types, but the levels of oxidative stress biomarkers were not significantly altered.

As we mentioned above, there is no official guideline for *in vitro* comet assay; many protocols describe how to perform it. As a consequence there is great variability in the results, both intra- and inter-laboratory, which has been described and analysed in several studies (Kumaravel et al. 2009; Azqueta et al. 2011; Collins et al. 2014). In order to obtain our results in bovine/animal lymphocytes, we tested several *in vitro* methods and procedures at first, and the best experimental conditions were then chosen and set up.

First of all the total cell count and cell viability were measured using the Trypan blue exclusion method. Cell viability is expressed as the percentage of viable cells. Results of cell viability are shown in Figures 5, 6, 7 and 8. At all tested concentrations and conditions, the cell viability was over 83%. To avoid obtaining false positive results, it is advisable to use the highest treatment concentration where the cell viability is over 70% immediately after treatment (Sasaki et al. 2007; Tice et al. 2010; Koppen et al. 2017). Some researchers apply higher percentages of cell viability in their experiments, e.g., over 75% (Giannotti et al. 2002; Kimura et al. 2013) or above 80 % (Bianchi et al. 2015; Manzano et al. 2015).

The exposure time is also a very important factor which can influence the final results of comet assay. Long-term exposure is not suitable for chemicals which are unstable in the culture media. Individual incubation times should be determined with them in mind. The first reason is that the initial DNA damage would be repaired, and secondly that there would not be a damage because of inactivation of the tested chemical. This process may lead to false negative results (Sekihashi et al. 2003; Sasaki et al. 2007). The initial repair process lasts about 15 min (more than 50% of damage), and then between 14 – 24% of the damage remains unrepaired after 60 min (Alapetite et al. 1999). Moreover, the DNA repair process could be inhibited by exposure to chemicals which suppress it. Cells finally stop growing, or go into cell-cycle arrest (Smith and Fornace 1996; Zhang et al. 2019). According to Tice et al. (2000), an appropriate exposure time for chemical *in vitro* genotoxicity assessment should be around 3 to 6 h in the presence or absence of a metabolic activation system. On the other hand, published results show that different exposure times are used, e.g., 1 h, 4 h, or 24 h (Lebaily et al. 1997; Zeljezic et al. 2016), and up to the 72 h (Celik et al. 2014) in the presence or absence of a mitotic activator. All our experiments were performed after 2-hour exposure to the tested insecticides, and this treatment time was successfully used in genotoxicity testing in several other studies (Chye et al. 2007; Magdolenova et al. 2014; Shaik et al. 2016).

Our comparison of different culture conditions, with (proliferating cells) and without (non-proliferating) mitotic activator (ANOVA and Dunnet's aposteriori test), showed that commercial formulation Calypso® was able to cause significant changes while pure agent thiacloprid was not, in both alkaline ($p < 0.05$, $p < 0.01$, $p < 0.001$; Figure 2) and neutral ($p < 0.05$; Figure 4) versions of comet assay. The reason for this may lie in the different mechanisms of detection; the initial genome damage may be repaired during the cultivation process. Comet assay is indicative only for initial genome damage when the repair process is insufficient or not allowed to occur (Kassie et al. 2000). Secondly, chemical adjuvants present in the commercial pesticide formulation may have a more harmful effect than the pure active agents (Chan et al. 2007; Cavas and Konen 2007).

In conclusion, commercial insecticide Calypso® can alter genetic material, and comet assay was found to be appropriate for the early detection of DNA damage in animal cell cultures under various cell culture conditions.

ACKNOWLEDGMENTS

This work was supported by IGA 09/2017: Detection of DNA damage in lymphocytes after exposure pesticides and the Slovak Scientific Agency VEGA 1/0242/19.

REFERENCES

Afanasieva, Katerina. & Andrei, Sivolob. (2018). "Physical principles and new applications of comet assay." *Biophysical Chemistry*, 238, 1-7.

Alapetite, Claire., Pierre, Thirion., Anne, de la Rochefordière., Jean-Marc, Cosset. & Ethel, Moustacchi. (1999). *International Journal of Cancer* 83, 83-90.

Al-Sarar, Ali S., Yasser, Abobakr., Alaa, E. Bayoumi. & Hamdy, I. Hussein. (2015). "Cytotoxic and genotoxic effects of abamectin, chlorfenapyr, and imidacloprid on CHO K1 cells." *Environmental Science and Pollution Research*, 22, 17041-17052.

Azqueta, Amaya. & Andrew, R. Collins. (2013). "The essential comet assay: a comprehensive guide to measuring DNA damage and repair." *Archives of Toxicology*, 87, 949-968.

Azqueta, Amaya., Kristine, B. Gutzkow., Gunnar, Brunborg. & Andrew, R Collins. (2011). "Towards a more reliable comet assay: Optimising agarose concentration, unwinding time and electrophoresis conditions." *Mutation Research/Genetic Toxicology and Environmental Mutagenesis*, 724, 41-45.

Bajpayee, Mahima., Ashutosh, Kumar. & Alok, Dhawan. (2013). "The comet assay: assessment of *in vitro* and *in vivo* damage." In *Genotoxicity assessment: methods and protocols*, edited by Alok

Dhawan and Mahima Bajpayee, 325-345. New York: Springer Science and Business Media.

Berheim, Elise H., Jonathan, A. Jenks., Jonathan, G. Lundgren., Eric, S. Michel., Daniel, Grove. & William, F. Jensen. (2019). "Effects of neonicotinoid insecticides on physiology and reproductive characteristics of captive female and fawn white-tailed deer." *Scientific Reports*, *9*, 4534.

Bernard, Fabien., Franck, Brulle., Sylvain, Dumez., Sébastien, Lemiere., Anne, Platel., Fabrice, Nesslany., Damien, Cuny,. Annabelle, Deram. & Franck, Vandenbulcke. (2014). "Antioxidant responses of Annelids, Brassicaceae and Fabaceae to pollutants: a review." *Ecotoxicology and Environmental Safety*, *114*, 273-303.

Bianchi, Jaqueline., Diogo, Cavalcanti Cabral-de-Mello., Maria, Aparecida. & Marin, Morales. (2015). "Toxicogenetic effects of low concentrations of the pesticides imidacloprid and sulfentrazone individually and in combination in *in vitro* tests with HepG2 cells and Salmonella typhimurium." *Ecotoxicology and Environmental Safety*, *120*, 174-183.

Blacquière, Tjeerd., Guy, Smagghe., Cornelis, van Gestel. & Veerle, Mommaerts. (2012). "Neonicotinoids in bees: a review on concentrations, side-effects and risk assessment." *Ecotoxicology*, *21*, 973–992.

Brandt, Annely., Anna, Gorenflo., Reinhold, Siede., Marina, Meixner. & Ralph, Büchler. (2016). "The neonicotinoids thiacloprid, imidacloprid, and clothianidin affect the immunocompetence of honey bees (Apis mellifera L.)." *Journal of Insect Physiology*, *86*, 40-47.

Caldeón-Segura, María Elena., Sandra, Gómez-Arroyo., Rafael, Villalobos-Pietrini., Carmen, Martínez-Valenzuela., Yolanda, Carbajal-López., María, del Carmen Calderón-Ezquerro., Josefina, Cortés-Eslava., Rocío, García-Martínez., Diana, Flores-Ramírez., María, Isabel Rodríguez-Romero., Patricia, Méndez-Pérez. & Enrique, Banuelos-Ruíz. (2012). "Evaluation of genotoxic and cytotoxic effects in human peripheral blood lymphocytes exposed *in vitro* to neonicotinoid insecticides news." *Journal of Toxicology*, *612647*.

Cavas, Tolga., Nilüfer, Cinkılıc., Özgür, Vatan. & Dilek, Yılmaz. (2014). "Effects of fullerenol nanoparticles on acetamiprid induced cytoxicity and genotoxicity in cultured human lung fibroblasts." *Pesticide Biochemistry and Physiology*, *114*, 1-7.

Cavas, Tolga., Nilufer, Cinkılıc., Ozgur, Vatan., Dilek, Yılmaz. & Mumun, Coskun. (2012). "*In vitro* genotoxicity evaluation of acetamiprid in CaCo-2 cells usingthe micronucleus, comet and cH2AX foci assays." *Pesticide Biochemistry ad Physiology*, *104*, 212-217.

Cavas, Tolga. & Serpil, Konen. (2007). "Detection of cytogenetic and DNA damage in peripheral erythrocytes of goldfish (Carassius auratus) exposed to a glyphosate formulation using the micronucleus test and the comet assay." *Mutagenesis*, *22*, 263-268.

Celik, Ayla., Seda, Y. Ekinci., Gizem, Guler. & Seda, Yildirim. (2014). "*In vitro* genotoxicity of fipronil sister chromatid exchange, cytokinesis block micronucleus test, and comet assay." *DNA and Cell Biology*, *33*, 148-154.

Chan, Yin-Ching., Shih-Chieh, Chang., Shih-Ling, Hsuan., Maw-Sheng, Chien., Wei-Cheng, Lee., Jaw-Jou, Kang., Shun-Cheng, Wang S. & Jiunn-Wang, Liao. (2007). "Cardiovascular effects of herbicides and formulated adjuvants on isolated rat aorta and heart." *Toxicology in Vitro*, *21*, 595–603.

Chye, Soi Mei., You, Cheng Hseu., Shih-Hsiung, Liang., Chin-Hui, Chen., Ssu, Ching Chen. (2007). "Single strand DNA breaks in human lymphocytes exposed to *para*-phenylenediamine and its derivatives." *Bulletin of Environmental Contamination and Toxicology*, *80*, 58-62.

Collins, Andrew R., Amaia, Azqueta Oscoz., Gunnar, Brunborg., Isabel, Gaivão., Lisa, Giovannelli., Marcin, Kruszewski., Catherine, C. Smith. & Rudolf, Štětina. (2008). "The comet assay: topical issues." *Mutagenesis*, *23*, 143-151.

Collins, Andrew R., Naouale, El Yamani., Yolanda, Lorenzo., Sergey, Shaposhnikov., Gunnar, Brunborg. & Amaya, Azqueta. (2014). "Controlling variation in the comet assay." *Frontiers in Genetics*, *5*, 1-6.

Costa, Chiara., Virginia, Silvari., Antonietta, A. Melchini., Stefania, Catania., James, J. A. Hefron,, Ada, Trovato. & Rita, de Pasquale. (2009). "Genotoxicity of imidacloprid in relation to metabolic activation and composition of the commercial product." *Mutation Research/Genetic Toxicology and Environmental Mutagenesis*, *672*, 40-44.

Demsia, Georgia., Dimitris, Vlastos., Marina, Goumenou. & Demetrios, P. Matthopoulos. (2007). "Assessment of the genotoxicity of imidacloprid and metalaxyl in cultured human lymphocytes and rat bone-marrow." *Mutation Research/Genetic Toxicology and Environmental Mutagenesis*, *634*, 32-39.

De Souza, Andressa., Afonso, dos Reis Medeiros Ana Cláudia de Souza., Márcia, Wink., Ionara, Rodrigues Siqueira., Maria, Beatriz Cardoso Ferreira., Luciana, Fernandes., Maria, Paz Loayza Hidalgo. & Iraci, Lucena da Silva Torres. (2011). "Evaluation of the impact of exposure to pesticides on the health of the rural population: Vale do Taquari, State of Rio Grande do Sul (Brazil)." *Ciencia et Saude Coletiva*, *16*, 3519–3528.

Feng, Lei., Lan, Zhang., Yanning, Zhang., Pei, Zhang. & Hongyun, Jiang. (2015). "Inhibition and recovery of biomarkers of earthworm Eisenia fetida after exposure to thiacloprid." *Environmental Science and Pollution Research*, *22*, 9475-9482.

Feng, Shaolong., Zhiming, Kong., Xinming, Wan., Pingan, Peng. & Eddy, Y. Zeng. (2005). "Assessing the genotoxicity of imidacloprid and rh-5849 in human peripheral blood lymphocytes *in vitro* with comet assay and cytogenetic tests." *Ecotoxicology and Environmental Safety*, *61*, 239-246.

Fu, Dong-Jun., Ping, Li., Jiang, Song., Sai-Yang, Zhang. & Han-Zhong, Xie. (2019). "Mechanisms of synergistic neurotoxicity induced by two high risk pesticide residues – Chlorpyrifos and Carbofuran via oxidative stress." *Toxicology in Vitro*, *54*, 338-344.

Gábelová, Alena., Darina, Slameňová., Ľubica, Ružeková., Timea, Farkašová. & Eva, Horváthová. (1997). "Measurement of DNA strand breakage and DNA repair induced with hydrogen peroxide using single cell gel electrophoresis, alkaline DNA unwinding and alkaline elution of DNA." *Neoplasma, 44*, 380–388.

Galdíková, Martina., Katarína, Šiviková., Beáta, Holečková., Ján, Dianovský., Monika, Drážovská. & Viera, Schwarzbacherová. (2015). "The effect of thiacloprid formulation on DNA/chromosome damage and changes in GST activity in bovine peripheral lymphocytes." *Journal of Environmental Science and Health B, 50*, 698-707.

George, Jasmine. & Yogeshwer, Shukla. (2011). "Pesticides and cancer: insights into toxicoproteomic-based findings." *Journal of Proteomics, 74*, 2713-2722.

Giannnotti, E., Luca, Vandin., Paolo, Repeto. & Rodolfo, Comelli. (2002). "A comparison of the *in vitro* Comet assay with the *in vitro* chromosome aberration assay using whole human blood or Chinese hamster lung cells: validation study using a range of novel pharmaceuticals." *Mutagenesis, 17*, 163-170.

Gharsalli, Tarek. (2016). "Comet assay on toxicogenetics; several studies in recent years on several genotoxicological agents." *Journal of Environmental and Analytical Toxicology, 6*, 1-9.

Gollapudi, B. Bhaskar. & Gopala, Krishna. (2000). "Practical aspects of mutagenicity testing strategy: an industrial perspective." *Mutation Research, 455*, 21-28.

Gyori, M. Benjamin., Gireedhar, Venkatachalam., Thiagarajan, P. S., David, Hsu. & Marie-Veronique, Clement. (2014). "OpenComet: An automated tool for comet assay image analysis." *Redox Biology, 2*, 457–465.

Han, Wenchao., Ying, Tian. & Xiaoming, Shen. (2018). "Human exposure to neonicotinoid insecticides and the evaluation of their potential toxicity: An overview." *Chemosphere, 192*, 59-65.

Holečková, Beáta., Katarína, Šiviková., Ján, Dianovský., Martina, Galdíková. (2013)." Effect of triazole pesticide formulation on bovine culture cells." *Journal of Environmental Science and Health, Part B: Pesticides, Food Contaminants, and Agricultural Wastes, 48*, 1080-1088.

Horváthová, Eva., Monika, Šramková., Juraj, Lábaj. & Darina, Slameňová. (2006). "Study of cytotoxic, genotoxic and DNA-protective effects of selected plant essential oils on human cells cultured *in vitro*." *Neuroendocrinology Letters*, *27*, (Suppl.2), 44–47.

Iwasa, Takao., Motoyama, Naoki., John, T. Ambrose. & Michael Roe, R. (2004). "Mechanism for the different toxicity of neonicotinoid insecticides in the honey bee, Apis mellifera." *Crop Protection*, *23*, 371–378.

Jandacek, Ronald J. & Patrick, Tso. (2001). "Factors affecting the storage and excretion of toxic lipophilic xenobiotics." *Lipids*, *36*, 1289–1305.

Jeschke, Peter., Ralf, Nauen., Michael, Schindler. & Alfred, Elbert. (2011). "Overview of the status and global strategy for neonicotinoids." *Journal of Agricultural and Food Chemistry*, *59*, 2897-2908.

Kalantzi, Olga I., Ruth, E. Alcock., Paul, A. Johnston., David, Santillo., Ruth, L. Stringer., Gareth, O. Thomas. & Kevin, C. Jones. (2001). "The global distribution of PCBs and organchlorine pesticides in butter." *Environmental Science and Technology*, *35*, 1013–1018.

Karabay, N. Ulku. & Gunnehir Oguz, M. (2005). "Cytogenetic and genotoxic effects of the insecticides, imidacloprid and methamidophos." *Genetics and Molecular Research*, *4*, 653-662.

Kassie, Fekadu., Wolfram, Parzefall. & Siegfried, Knasmuller. (2000). "Single cell gel electrophoresis assay: a new technique for human biomonitoring studies." *Mutation Research*, *463*, 13–31.

Kimura, Aoi., Atsuro, Miyata. & Masamitsu, Honma. (2013). "A combination of *in vitro* comet assay and micronucleus test using human lymphoblastoid TK6 cells." *Mutagenesis*, *28*, 583-590.

Końca, Krzysztof., Anna, Lankoff., Anna, Banasik., Halina, Lisowska., Tomasz, Kuszewski., Stanisław, Góźdź., Zbigniew, Koza. & Andrzej, Wojcik. (2003). "A cross platform public domain PC image analysis program for the comet assay." *Mutation Research*, *534*, 15-20.

Koppen, Gudrun., Amaya, Azqueta., Bertrand, Pourrut., Gunnar, Brunborg., Andrew, R. Collins. & Sabine, A. S. Langie. (2017). "The next three decades of the comet assay: a report of the 11th International Comet Assay Workshop." *Mutagenesis*, *32*, 397-408.

Kuchařová, Monika., Miroslav, Hronek., Katerina, Rybáková., Zdenek, Zadák., Rudolf, Štětina., Vera, Josková. & Anna, Patková. (2019). "Comet assay and its use for evaluating oxidative DNA damage in some pathological states." *Physiological Research, 68*, 1-15.

Kumaravel, T. S., Barbara, Vilhar., Stephen, P. Faux. & Awadhesh, N. Jha. (2009). "Comet Assay measurements: a perspective." *Cell Biology and Toxicology, 25*, 53-64.

Langie, Sabine A. S., Amaya, Azqueta. & Andrew, R. Collins. (2015). "The comet assay: past, present, and future." *Frontiers in Genetics, 6*, 1-3.

Lebaily, Pierre., Carole, Vigreux., Thierry, Godard., Francois, Sichel., Edith, Bar J. Y., Le, Talaer., Michel, Henry-Amar. & Pascal, Gauduchon. (1997). "Assessment of DNA damage induced *in vitro* by etoposide and two fungicides (carbendazim and chlorothalonil) in human lymphocytes with the comet assay." *Mutation Research, 375*, 205-217.

Li, Bing., Xiaoming, Xia., Jinhua, Wang., Lusheng, Zhu.., Jun, Wang. & Guangchi, Wang. (2018). "Evaluation of acetamiprid-induced genotoxic and oxidative responses in Eisenia fetida." *Ecotoxicology and Environmental Safety, 161*, 610-615.

Lioi, Maria B., Maria, R. Scarfi., Antonietta, Santoro., Rocchina, Barbieri., Olga, Zeni., Dino, Di Berardino. & Matilde, V. Ursini. (1998). "Genotoxicity and oxidative stress induced by pesticide exposure in bovine lymphocyte cultures *in vitro*." *Mutation Research, 403*, 13–20.

Lioi, Maria B., Antonietta, Santoro., Rocchina, Barbieri., Salvatore, Salzano. & Matilde, V. Ursini. (2004). "Ochratoxin A and zearalenone: a comparative study on genotoxic effects and cell death induced in bovine lymphocytes." *Mutation Research/Genetic Toxicology and Environmental Mutagenesis, 557*, 19-27.

López, Silvia L., Delia, Aiassa., Stella, Benítez-Leite., Rafael, Lajmanovich., Fernando, Manas., Gisela, Poletta., Norma, Sánchez., María, Fernanda Simoniello. & Andrés, E. Carrasco. (2012). "Pesticides used in South American GMO-based agriculture: a review of their effects on humans and animal models." *Advances in Molecular Toxicology*, *6*, 41-75.

Lorenzo, Yolanda., Solange, Costa., Andrew, R. Collins. & Amaya, Azqueta. (2013). "The comet assay, DNA damage, DNA repair and cytotoxicity: hedgehogs are not always dead." *Mutagenesis*, *28*, 427-432.

Lundin, Ola., Maj, Rundlöf., Henrik, G. Smith., Ingemar, Fries. & Riccardo, Bommarco. (2015). "Neonicotinoid insecticides and their impacts on bees: a systematic review of research approaches and identification of knowledge gaps." *PLoS one*, *10*, e0136928.

Magdolenova, Zuzana., Martina, Drlickova., Kristi, Henjum., Elise, Rundén-Pran., Jana, Tulinska., Dagmar, Bilanicova., Giulio, Pojana., Alena, Kazimirova., Magdalena, Barancokova., Miroslava, Kuricova., Aurelia, Liskova., Marta, Staruchova., Fedor, Ciampor., Ivo, Vavra., Yolanda, Lorenzo., Andrew, Collins., Alessandra, Rinna., Lise, Fjellsbø., Katarina, Volkovova., Antonio, Marcomini., Mahmood, Amiry-Moghaddam. & Maria, Dusinska. (2015). "Coating-dependent induction of cytotoxicity and genotoxicity of ironoxide nanoparticles." *Nanotoxicology*, *9* (suppl. 1), 44-56.

Maienfisch, Peter., Hanspeter, Huerlimann., Alfred, Rindlisbacher., Laurenz, Gsell., Hansruedi, Dettwiler., Joerg, Haettenschwiler., Evelyne, Sieger. & Markus, Walti. (2001). "The discovery of thiamethoxam: a second-generation neonicotinoid." *Pest Management Science*, *57*, 165-176.

Manzano, Bárbara C., Matheus, M. Roberto., Márcia, M. Hoshina., Amauri, A. Menegário. & Maria, A. Marin-Morales. (2015). "Evaluation of the genotoxicity of waters impacted by domestic and industrial effluents of a highly industrialized region of São Paulo State, Brazil, by the comet assay in HTC cells." *Environmental Science and Pollution Research*, *22*, 1399–1407.

Moller, Peter. (2018). "The comet assay: ready for 30 more years." *Mutagenesis*, *33*, 1-7.

Morrissey, Christy A., Pierre, Mineau., James, H. Devries., Francisco, Sanchez-Bayo., Matthias, Liess., Michael, C. Cavallaro. & Karsten,

Liber. (2015). "Neonicotinoid contamination of global surface waters and associated risk to aquatic invertebrates: a review." *Environment International*, 74, 291–303.

Mostafalou, Sara. & Mohammad, Abdollahi. (2012). "The role of environmental pollution of pesticides in human diabetes." *International Journal of Pharmacology*, 8, 139–140.

Neri, Monica., Daniele, Milazzo., Donatella, Ugolini., Mirta, Milic., Alessandra, Campolongo., Patrizio, Pasqualetti. & Stefano, Bonassi. (2015). "Worldwide interest in the comet assay: a bibliometric study." *Mutagenesis*, 30, 155-163.

Panda, K. Santosh. & Balachandran, Ravindran. (2013). "Isolation of Human PBMCs." *Bio-protocol*, 3(3), e323. doi: 10.21769/BioProtoc.323.

Sasaki, Yu F., Takanori, Nakamura. & Satomi, Kawaguchi. (2007). "What is better experimental design for *in vitro* comet assay to detect chemical genotoxicity? *AATEX*, 14, special issue, 499-504.

Schwarzbacherová, Viera., Maciej, Wnuk., Anna, Lewinska., Leszek, Potocki., Jacek, Zebrowski., Marek, Koziorowski., Beáta, Holečková., Katarína Šiviková. & Ján, Dianovský. (2017). "Evaluation of cytotoxic and genotoxic activity of fungicide formulation Tango (R) Super in bovine lymphocytes." *Environmental Pollution*, 220, 255-263.

Sekihashi, Kaoru., Hiromi, Saitoh., Ayako, Saga., Kazushige, Hori., Munehiro, Nakagawa., Makoto, Miyagawa. & Yu, F. Sasaki. (2003). "Effect of *in vitro* exposure time on comet assay results." *Environmental Mutagen Research*, 25, 83-86.

Sekeroglu, Vedat., Zulal, Atli Sekerolu. & Haluk, Kefelioglu. (2013). "Cytogenetic effects of commercial formulations of deltamethrin and/or thiacloprid on Wistar rat bone marrow cells." *Environmental Toxicology*, 28, 524-531.

Sekeroglu, Zulal Atli., Vedat, Sekerolu., Ebru, Ugun., Seval, Kontas Yedier. & Birsen, Aydın. (2018). "Cytotoxicity and genotoxicity of clothianidin in human lymphocytes with or without metabolic activation system." *Drug and Chemical Toxicology*, 42, 364-370.

Shaika, Asma Sultana., Abjal, Pasha Shaikb. & Kaiser Jamilcand, Abbas H. Alsaeed. (2016). "Evaluation of cytotoxicity and genotoxicity of pesticide mixtures onlymphocytes." *Toxicology Mechanisms and Methods*, 26, 588-594.

Singh, Narendra P. (2000). "Microgels for estimation of DNA strand breaks, DNA protein crosslinks and apoptosis." *Mutation Research*, 455, 111–27.

Singh, Narendra P. (2016). "The comet assay: Reflections on its development, evolution and applications." *Mutation Research/Reviews in Mutation Research*, 767, 23-30.

Singh, P. Narendra., Michael, T. McCoy., Raymond, R. Tice. & Edward, L. Schneider. (1988). "A simple technique for quantitation of low levels of DNA damage in individual cells." *Experimental Cell Research*, 175, 184–91.

Slameňová, Darina., Eva, Horváthová., Soňa, Robichová., Ľubica, Hrušovská., Alena, Gábelová., Karol, Kleibl., Jana, Jakubíková. & Ján, Sedlák. (2003). "Detection of MNNG-induced DNA lesions in mammalian cells; validation of comet assay against DNA unwinding technique, alkaline elution of DNA." *Environmental and Molecular Mutagenesis*, 41, 28–36.

Smith, Martin L. & Albert, J. Fornace. Jr. (1996). "Mammalian DNA damage-inducible genes associated with growth arrest and apoptosis." *Mutation Research*, 340, 109-124.

Stivaktakis, Polychronis., Dimitris, Vlastos., Evangelos, Giannakopoulos. & Demetrious, P. Matthopoulos. (2010). "Differential micronuclei induction in human lymphocyte cultures by imidacloprid in the presence of potassium nitrate." *The Scientific World Journal*, 10, 80-89.

Tice, Raymond R., Eva, Agurell., Diana, Anderson., Brian, Burlinson., Andreas, Hartmann., Hiroshi, Kobayashi., Youichi, Miyamae., Emilio, Rojas., Jae-Chun, Ryu. & Yu, F. Sasaki. (2000). "The single cell gell/comet assay: guidelines for *in vitro* and *in vivo* genetic toxicology testing." *Environmental and Molecular Mutagenesis*, 35, 206-221.

Tomizawa, Motohiro. & John, E. Casida. (2005). "Neonicotinoid insecticide toxicology: mechanisms of selective action." *Annual Review of Pharmacology and Toxicology*, *54*, 247-68.

Willett, Lyn B., O'Donnell, A. F., Holly, I. Durst. & Kurz, M. M. (1993). "Mechanisms of movement of organochlorine pesticides from soils to cows via forages." *Journal of Dairy Science*, *76*, 1635–1644.

Williams, Geoffrey R., Aline, Troxler., Gina, Retschnig., Kaspar, Roth., Orlando, Yañez., Dave, Shutler., Peter, Neumann. & Laurent, Gauthier. (2015). "Neonicotinoid pesticides severely affect honey bee queens." *Scientific reports*, *5*, 14621.

Wood, Thomas J. & Dave, Goulson. (2017). "The environmental risks of neonicotinoid pesticides: a review of the evidence post 2013." *Environmental Science and Pollution Research*, *24*, 17285-17325.

Zeljezic, Davor., Marin, Mladinic., Suzana, Zunec., Ana Lucic, Vrdoljak., Vilena, Kasuba., Blanka, Tariba., Tanja, Zivkovic., Ana, Marija Marjanovic., Ivan, Pavicic., Mirta, Milic., Ruzica, Rozgaj. & Nevenka, Kopjar. (2016). "Cytotoxic, genotoxic and biochemical markers of insecticide toxicity evaluated in human peripheral blood lymphocytes and an HepG2 cell line." *Food and Chemical Toxicology*, *96*, 90-106.

Zhang, Wei., Ruiguo, Wang., John, P. Giesy., Yang, Li. & Peilong, Wang. (2019). "Tris (1,3-dichloro-2-propyl) phosphate treatment induces DNA damage, cell cycle arrest and apoptosis in murine RAW264.7 macrophages." *Journal of Toxicological Sciences*, *44*, 133-144.

In: A Closer Look at the Comet Assay ISBN: 978-1-53611-028-9
Editor: Keith H. Harmon © 2019 Nova Science Publishers, Inc.

Chapter 7

DETERMINATION OF ALUMINUM-INDUCED OXIDATIVE AND GENOTOXIC EFFECTS IN SUNFLOWER LEAVES

Aslıhan Çetinbaş-Genç[1], Elif Kılıç-Çakmak[2], Fatma Yanık[1], Filiz Vardar[1,*], Ahu Altınkut-Uncuoğlu[2] and Yıldız Aydın[1]

[1]Marmara University, Science and Arts Faculty,
Department of Biology, Istanbul, Turkey
[2]Marmara University, Engineering Faculty,
Department of Bioengineering, Istanbul, Turkey

ABSTRACT

According to rapid industrialization and urbanization serious ecological problems come into prominence worldwide. Agronomic crops and wild flora are faced with environmental risk factors primarily inducing reduced crop quality and yield due to their sessile nature. Most of the economic losses are related to soil acidification and heavy metal accumulation causing adverse effects on plant growth and development. Aluminum (Al) is one of the most abundant elements in the earth crust

---

[*] Corresponding Author's Email: filiz.vardar@gmail.com.

comprising 8.1%. It exists in the form of insoluble aluminosilicates or oxides commonly. If the soil pH reduces (pH <5.0), the complex Al is dissolved and absorbed by plant root system which is the first target organ in plants. Since the root apex accumulates more Al, Al toxicity represents its adverse effects on root growth initially resulting in low water and nutrient uptake. Because of being the first target organ, most of the Al toxicity researches subjected root growth and development, whereas limited studies are available on foliar symptoms of Al toxicity. In the present study *Helianthus annuus* (sunflower) seedlings were irrigated with Hoagland solution containing with or without different concentrations of $AlCl_3$ (50, 100, 150 and 200 µM pH 4.5). After eight weeks, fresh leaves were analyzed for determining the oxidative and genotoxic responses. Morphological parameters such as germination rate and stoma number were evaluated. Antioxidant enzyme analyses (superoxide dismutase, catalase and peroxidase activity) were performed in sunflower leaves for determination of oxidative stress. Total chlorophyll, carotenoid and anthocyanine content were also measured. The genotoxic effects of Al were performed by comet assay in sunflower leaves. It has been known that comet assay is a practical and sensitive implementation to assess the genotoxic impact of various type of stress factors such as pesticides, UV and heavy metals. The assay is based on the quantification of denatured DNA fragments which migrates out of the nucleus during electrophoresis. According to our results, Al caused adverse effects on sunflower leaves in all concentrations. Although the toxicity level was dose dependent up to 150 µM, it reduced at 200 µM in compare to 150 µM. The comet assay results also revealed that Al induced DNA damage confirmed by increase in % DNA tail, olive tail moment and tail intensity. The DNA damage was evident in sunflower leaves which is not the first target as roots. In conclusion, sunflower leaves showed a resistance up to 150 µM Al by enhancing its stress tolerance mechanism, but 200 µM Al was over dose that blocked its balance.

Keywords: aluminum toxicity, comet assay, leaves, oxidative stress, sunflower

INTRODUCTION

Aluminum (Al) is one of the most abundant and common elements in the earth crust comprising 8.1% (Li et al., 2016). Whereas it is very common in soil, it is not an essential element for living organisms as well as plants (Matsumoto, 2000; Abate et al., 2013). Al is found as non-toxic complexes in the neutral or weakly acidic soils as insoluble aluminosilicates or oxides; however multiple natural or environmental factors cause to reduce the soil acidity and to solubilize the complex Al into phytotoxic forms. $Al(H_2O)_6^{3+}$ is the most toxic and prevalent compound in low pH soils and dissolves to Al^{3+} ions which is highly reactive. Thus, Al is regarded as an abiotic stress factor under pH < 5 for all living organisms (Shaw and Tomljenovic, 2013; Yamamoto, 2018).

Al is one of the major plant growth-limiting factors in acid soils. The estimations indicate that approximately 70% of the world's potential farmable lands have acid soil and plants are face to Al toxicity (Abate et al. 2013; Ma et al. 2014). Natural acidic soil occurrence is widespread in tropical and subtropical zones such as South America, Central Africa and Southeast Asia. Besides, acidic soil formation is also a critical environmental problem in temperate zones including USA, Canada and Europe countries in which industrial activities are very intensive. In these regions environmental pollution and acid rains are prominent which raise the solubilized Al^{3+} ions in the soil (Vitorello et al., 2005). Moreover, exhaustive agriculture and inappropriate farming, over leaching with basic actions are also enhanced the soil acidity resulting in increased toxic (Al, Cd and Mn) and reduced essential minerals (P, Ca, Mg and Mo) (Vardar et al., 2018).

Several researches revealed the adverse effects of Al ions on plant growth and development within a few minutes even at low micro-molar doses in acid soils. Therefore, Al ions are regarded as a primary constraint for crop production. Although the first target organ is root, toxic Al ions have also adverse effects on the other organs' structure and anatomy. Regarding the whole plant; reduction in total leaf number, regressed photosynthesis, chlorosis, necrosis, reduction of stomatal aperture and number, reduction in root-shoot biomass, inhibition of root growth are the common response to Al stress (Vardar and Ünal, 2007; Abate et al., 2013; 2015).

Well-documented researches prove that the root apex (root cap-meristem-elongation zone) as a first target plays a critical role in Al uptake. As a result of excess Al uptake, the normal root apices change to swollen, cracked, stubby and stiff appearance (Matsumoto, 2000; Vardar et al., 2006). Besides, regression in fine branching and root hairs are also evident (Ciamporova, 2002). Moreover, Al-induced root growth inhibition causes low uptake of soil water and mineral elements such as Ca, Mg, K, Fe and P resulting in reduced yield and crop quality (Delhaize et al., 2004; Singh et al., 2017). Inhibition of root growth is considered as the consequence of interfered lateral and longitudinal cell growth, prior to arrested cellular division (Frantzios et al., 2001; Ciamporova, 2002; Vardar and Ünal, 2007).

Many researchers described Al-induced cellular alterations in roots including enhancement in cell wall rigidity by cross-linking pectin, lignin accumulation, induction of callose synthesis, reduction in starch grains, lipid peroxidation, disruption of ion transport, breakdown of Ca^{2+} homeostasis, structural alterations in cytoskeleton, excessive generation of reactive oxygen species, and alterations in antioxidant enzyme activities. Well-documented studies indicate that Al has a strong binding affinity to oxygen donor ligands, such as polypeptides, phospholipids, flavonoids, anthocyanin, carboxylic acids, inorganic phosphate, DNA and RNA. As a result of the binding affinity of Al^{3+}, the function and integrity of the cells are affected severely (Matsumoto, 2000; Delhaize at al., 2004; Poschenrieder et al., 2008; Vardar et al., 2011; 2018).

Existing researches concern that Al has a genotoxic profile and Al exposure can lead adverse effects on DNA composition and replication based on forming more rigid double helix and chromatin structure. It has been known that Al reduces mitotic index and increases abnormal chromosome movements which are also related to Al-induced alterations in tubulin polymerization and depolymerization (Silva et al., 2000; Frantzios et al., 2000; Vardar et al., 2011).

Besides, recent studies reveal that Al causes programmed cell death (PCD) resulting in DNA fragmentation. It is one of the important signatures of PCD originates from internucleosomal DNA cleavage by specific proteases and nucleases. DNA fragmentation can also be analyzed by TUNEL reaction, comet assay, laddering on the agarose gel and flow cytometry (Tripathi et al., 2016). Vardar et al. (2015; 2016) revealed DNA fragmentation using comet assay and agarose gel electrophoresis under Al stress in early hours in wheat, rye, barley, oat and triticale.

The results up to date reveal that plant genotype, cell/tissue type, pH, duration, concentration and content of external solution (types of chelators, concentrations of Ca^{2+} and other cations) are decisive on the severity of Al toxicity (Kinraide and Parker, 1987).

Although many researches were conducted to clarify the Al toxicity mechanism and tolerance based on root system, the other plant organs such as leaves are of secondary importance. The detailed studies performed on whole plant systems will help to improve our understanding on Al toxicity mechanism with the goal in mind for environmental monitoring. Thus, we designed a long-term study with different concentrations of $AlCl_3$ to reveal the responses of sunflower leaves to Al toxicity.

MATERIAL AND METHODS

The seeds of *Helianthus annuus* L. (sunflower) var. AGA-1301 obtained from AGROMAR (Bursa, Turkey) were sterilized with 1% sodium hypochlorite (NaOCl) solution through a magnetic stirrer. After rinsing with distilled water, peeled seeds were germinated in petri dishes with Hoagland solution (pH 6.0) with or without $AlCl_3$ (50, 100, 150 and 200 μM pH 4.5). Processes were performed in plant growth room with irradiance of 5000 lx (day/night 16:8, respectively), temperature of 23 ± 2°C, and relative humidity of 45 – 50%. The Hoagland solution includes 5 mM $Ca(NO_3)_2$, 5 mM KNO_3, 2 mM $MgSO_4$, 1 mM KH_2PO_4, 30 μM Fe(III)-EDTA and standard Hoagland micronutrients (Hoagland and Arnon, 1950). Seed germination (SG) and relative seed germination (RSG) were calculated according to the formula (Jalili et al., 2019):

$$SG = (\text{germinated seed number/total seed number}) \times 100$$
$$RSG = [\text{germinated seed number (sample)/germinated seed number (control)}] \times 100$$

The sunflower seedlings were transferred to pots and irrigated with solutions at 2-day intervals after germination. After 8 weeks the plant leaves harvested and used for further analyses.

Stoma numbers which localize in abaxial surface were counted from fresh leaves of control and experimental groups for anatomical observations (Radoglou and Jarvis, 1990). Photosynthetic pigment determination (chlorophyll a, b, chlorophyll a/b, total chlorophyll and carotenoids) was performed according to Arnon (1949). Fresh leaves were homogenized in 80% ice-cold acetone with chilled mortar and pestle. The homogenates were centrifuged at 3000 g for 10 min at + 4°C. The volume of supernatants was measured in dark at 470, 645 and 663 nm, spectrophotometrically. The photosynthetic pigment concentrations were calculated according to the formula below, and expressed as mg/ml.

Chlorophyll a = [(12.7 × A_{663} − 2.69 × A_{645}) × final volume]/1000
Chlorophyll b = [(22.9 × A_{645} − 4.68 × A_{663}) × final volume]/1000
Total Chlorophyll = [(20.2 × A_{645} + 8.02 × A_{663}) × final volume]/1000

Control and Al treated leaves (100 mg) were homogenized with 1 ml of cold sodium-phosphate buffer (PBS, 50 mM, pH 7.0) with chilled mortar and pestle for antioxidant enzyme activities. The leaf homogenates were centrifuged at 14000 rpm for 20 min at + 4°C. The supernatant was transferred to new microcentrifuge tubes and stored in ice for superoxide dismutase (SOD), catalase (CAT) and guaiacol peroxidase (POD) enzyme activity assays.

The activity of SOD was evaluated by Cakmak and Marschner (1992). The reaction mixture containing 2 ml of substrate buffer (0.1 M PBS, pH 7.0; 2 M Na_2CO_3; 0.5 M EDTA; 0.3 M L-methionin; 7.5 mM NBT; 0.2 mM riboflavin) and 2 µl of the supernatant was incubated under 15 W fluorescent lamps for 10 min, and measured immediately at 560 nm spectrophotometrically. One unit of SOD is determined as the amount required inhibiting the photo reduction of NBT by 50%. The results were expressed in term of SOD unit(U)/mg protein/time.

The activity of CAT was analyzed as described by Cho et al. (2000). The reaction mixture containing 1 ml of substrate buffer (20 mM PBS, pH 7.0; 6 mM H_2O_2) and 25 µl of enzyme extract was measured by the decrease in absorbance for 2 min at 240 nm, spectrophotometrically. The results were expressed as ΔA/min/mg protein.

The activity of POD was determined by the method of Birecka et al. (1973). The reaction mixture containing 1.5 ml of substrate buffer (0.1 M PBS, pH 5.8; 5 mM H_2O_2; 15 mM guaiacol) and 10 µl of enzyme extract was measured immediately for 2 min at 470 nm, spectrophotometrically. The results were expressed as ΔA/min/mg protein.

Determination of DNA damage was performed by comet assay (alkaline single cell gel electrophoresis) method described by Jouvtchev et al. (2001) with minor modifications. Hoagland without $AlCl_3$ was used for negative control and Hoagland with 0.1% H_2O_2 used for positive control group. The nuclei were isolated from control and Al treated sunflower leaves (0.4 g) by careful slicing with a razor blade in 0.4 ml Tris-HCl buffer (0.4 M, pH 7.5) on ice, in dark conditions. The homogenate filtered through sterile 20 µm filter. For each group, 100 µl of the nuclei suspension were mixed with 0.8 ml hot 0.65% low-melting point agarose (LMA) in a micro-centrifuge tube, and dropped onto slides pre-coated with 0.65% normal melting point agarose (HMA). Each drop was covered with a 24 × 60 mm cover slip and slides were stored at +4°C for 10 min. After removal of the cover slips, the HMA-coated slides were placed submerged in alkaline buffer (0.3 M NaOH; 1 mM EDTA) for 15 min that facilitated nuclear DNA unwinding, followed by electrophoresis at 26 V/300 mA for 30 min, under cold and dark conditions. After electrophoresis, the slides were neutralized in 0.1 M PBS, three times for 5 min. Slides were stained by 0.2% ethidium bromide (EB). Samples were examined with a Leica DM LB2 fluorescence microscope, equipped with a 515-560 nm excitation filter and a 590 nm barrier filter for EB. The Comet module of the KAMERAM image processing, and analyzing software were used for measurement of DNA content in the head and DNA tail of each comet. At least 2 slides from each independent experimental group were evaluated, with each providing a median value of %DNA in tail, olive tail moment and tail intensity of at least 50 comets.

All experiments were carried out at least three times. For each repetition and experimental groups, thirty seeds were germinated. Variance analysis of whole experimental data was performed with SPSS 16.0 statistical computer program. ANOVA (One way analyses of variance) with Tukey's posthoc HSD were applied. The significance of mean differences was at P < 0.05 levels.

RESULTS AND DISCUSSION

It has been widely known that roots are the first target of Al toxicity. Thus, most of the toxicity studies have been built on root - Al interaction (Vardar et al., 2015; 2016). Besides, information concerning the responses to Al toxicity of leaves is insufficient. From this point of view, we designed a long-term study to evaluate dose-dependent effects of Al toxicity in sunflower leaves.

The seed germination is the critical initial stage of plant growth and development. The successful germination process contains 3 basic stages: (1) imbibition, (2) radicle emergence and (3) radicle elongation. The last stage radicle elongation specifies the germination success of the seeds (Weitbrecht et al., 2011). To analyze the seed germination rate is one of the effective and economical bioassays to calculate the potential of Al toxicity in plants (Luo et al., 2018).

Based on our results, the seed germination rates were 98% in control, 97% in 50 µM, 95% in 100 µM, 93% in 150 µM and 88% in 200 µM Al exposure after 72 h (Data not shown). The reduction in relative seed germination was 0.99% in 50 µM, 0.97% in 100 µM, 0.95% in 150 µM and 0.88% in 200 µM with respect to controls. The germination rates revealed that Al treatment reduced the seed germination dose dependently. The highest regression was observed in 200 µM Al treatment.

Studies concerning seed germination under Al toxicity are limited. Gumze et al. (2007) reported reduced germination rate in maize (*Zea mays*) after 50 µM Al treatment. Similarly, Al decreased germination of pea (*Pisum sativum*) seeds significantly (Singh et al., 2011). However, Vardar et al. (2006) notified delay, not inhibition, in germination of tobacco seeds with increasing Al concentrations (50 - 200 µM). On the contrary, Jamal et al. (2006) observed no significant effect on seed germination of wheat (*Triticum aestivum*).

Whereas there are no detailed reports on the mechanism of Al-induced inhibition of seed germination, Samad et al., (2017) elaborated a study concerning the reason of inhibition after Al exposure. The researchers observed reduction of germination in rice (*Oryza sativa*) and chickpea (*Cicer arietinum*) after 10, 50, 100 and 150 µM Al exposure at 48 h. However, at 96 h, the reduction was only visible between 50 - 150 µM. The researchers analyzed K^+ and Cl^- uptake and accumulation revealing K^+ reduction and Cl^- elevation in plumula and radicula after Al exposure. They concluded that, Al-induced breakdown of ion homeostasis may be responsible for delay and inhibition of seed germination.

Stomata have a key role in determining water use efficiency through the regulation of the exchange of H_2O and CO_2 between plant tissues and the atmosphere. It has been reported that abiotic stress such as heavy metals decrease stoma number and their aperture size inducing inhibition of transpiration (Shen et al., 2015). Therefore, stomatal parameters such as stomatal density, size, potential conductance and closure also can be used as Al toxicity and tolerance indicators (Smirnov et al., 2014)

For better comprehension of morphological-anatomical effects after Al exposure, stomata in the abaxial epidermis of sunflower leaves were counted. The stoma numbers reduced by 13.87% in 50 µM, 31.69% in 100 µM, 37.63% in 150 µM and 30.71% in 200 µM Al exposure in compare to control (Figure 1). The highest reduction was observed in 150 µM Al exposure.

Similar to our results, Smirnov et al. (2014) reported that 50 µM Al has adverse effects on stomatal parameters of *Fagopyrum esculentum* leaves in both abaxial and adaxial surfaces. Their results revealed that stomatal density (SD), stomatal index (SI) and stomatal shape coefficient (SSC) were significantly decreased after 10 days Al exposure. The researchers also included that the highest reduction in SD, SI and SSC parameters was at abaxial leaf surface. They also observed maximum inhibition of stomatal surface (μm²) at adaxial side. Smirnow et al. (2014) suggested that the reduction of stoma number could be a consequence of adverse effects of Al on protodermal sister cell division into guard cells.

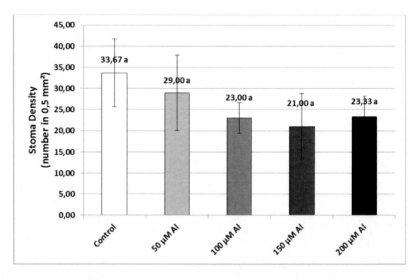

Figure 1. Stoma density of sunflower leaves in abaxial epidermis after Al exposure. Different letters concern significance of independent samples at P < 0.05 level. ± SD: standard deviation.

Although there are limited studies concerning Al-induced alterations on stomatal number, several studies reveal that heavy metal toxicity also declines stomatal number as in Al toxicity (Rucińska-Sobkowiak, 2016). Stomatal reduction after Cd (Kasim, 2006), Cu (Kasim, 2006), Zn (Kasim, 2007; Sagardoy et al., 2010) and As (Gupta and Bhatnagar, 2015) was also observed in several plant species. The researchers also documented that after heavy metal induction the quantity of abnormal (non-functional) stomata significantly increased (Rucińska-Sobkowiak, 2016). Gupta and Bhatnagar (2015) also suggested that disruption of microtubules may cause alterations of the cell division scheme resulted in abnormal stoma formation and arrested stomatal development. It has been widely known that Al also alters cell cytoskeleton (Matsumoto, 2000) and the reason of the reduction in stoma number may be the alterations in cytoskeleton.

In plants, photosynthesis is a very vital process during normal growth and development. For determination of photosynthetic activity; gaseous exchange measurements, determination of chlorophyll fluorescence and chlorophyll content are useful methods (KrishnaRaj et al., 2000; MacFarlane, 2003). Measurement of total chlorophyll content is also a useful assay to identify oxidative stress in green plants.

In sunflower leaves, total chlorophyll content increased in all of the Al concentrations with regard to control. It was increased by 29.6% in 50 μM, 60.2% in 100 μM, 120.4% in 150 μM and 65.3% in 200 μM. Even if total chlorophyll content increased in compare to control at 200 μM, the highest enhancement was observed at 150μM. In a similar way chl *a* and *b* content increased in compare to control (Table 1).

Shahnawaz et al. (2017) treated barley with different concentrations of Al (2 mM, 4 mM and 6 mM) for 6 days. According to their results chlorophyll *a*, *b* and total chlorophyll content reduced dose dependently. Similar results concerning Al toxicity on photosynthetic pigments were reported in *Cucumis sativus* (Pereira et al., 2006), *Citrus* (Jiang et al., 2008) and *Brassica napus* (Tohidi et al., 2015).

Most of the researches revealed that heavy metals including Al reduce the photosynthetic capacity of plants according to regression of chlorophyll content in many species (Zhang et al., 2007; Jiang et al., 2008; Radić et al., 2010; Karimaei and Poozesh, 2016). It has been known that Al^{3+} ions disable the activities of many cellular enzymes as well as in chloroplast. Delta-aminolevulinic acid dehydratase (ALAD) which is responsible for chlorophyll synthesis is restrained by Al^{3+} ions. During ALAD synthesis Al^{3+} compete with Mg^{2+} for connecting to active site of enzyme leading to inhibition of chlorophyll synthesis (Pereira et al., 2006; Haider et al., 2007).

On the contrary, we observed dose-dependent increase of chlorophyll in sunflower leaves after Al exposure. However, a reduction was observed in 200 μM in compare to lower concentrations suggesting an adaptive response. It can be suggested that higher concentrations from 200 µM may decrease chlorophyll content as described in previous studies.

Carotenoids are ubiquitous and essential pigments in plants that responsible for capturing energy, not absorbing light. They also function as non-enzymatic antioxidants protecting the chlorophyll pigments under stressful conditions (Strzalka et al., 2003). In a general opinion heavy metal induction reduces both chlorophyll and carotenoid contents (Baek et al., 2012), but according to the metal type, dose, exposure time and plant species carotenoid content may enhance (Sinha et al., 2003). Anthocyanins are accessory pigments which are not directly involved in photosynthesis. In response to heavy metal stress anthocyanins are also produced to protect plants as non-enzymatic antioxidants from various types of stresses (Hale et al., 2001; Neill et al., 2002).

According to our results, carotenoid content raised as it was in total chlorophyll. It was increased by 16.7% in 50 µM, 46.7% in 100 µM, 116.7% in 150 µM and 70% in 200 µM. The highest increase was at 150 µM. However, anthocyanin content exhibited a steady increase. It was increased by 2.6% in 50 µM, 1.9% in 100 µM, 11.2% in 150 µM and 20% in 200 µM (Table 1).

Table 1. Chlorophyll, carotenoid and anthocyanin contents of sunflower leaves after Al exposure. Different letters concern significance of independent samples at P < 0.05 level.
± SD: standard deviation

	Chl *a* (mg/ml)	Chl *b* (mg/ml)	Chl a/b	Total Chl. (mg/ml)	Carotenoids (mg/ml)	Anthocyanin (mg/ml)
Control	0.062[a] ± 0.02	0.036[a] ± 0.01	1.72	0.098[a] ± 0.04	0.030[a] ± 0.01	7.160[a] ± 0.04
50 µM Al	0.082[a] ± 0.01	0.045[a] ± 0.01	1.82	0.127[a] ± 0.01	0.035[ab] ± 0.01	7.347[a] ± 1.49
100 µM Al	0.100[a] ± 0.01	0.058[a] ± 0.01	1.72	0.157[a] ± 0.03	0.044[ab] ± 0.01	7.296[a] ± 0.73
150 µM Al	0.131[a] ± 0.05	0.085[a] ± 0.04	1.54	0.216[a] ± 0.08	0.065[b] ± 0.02	7.960[a] ± 0.36
200 µM Al	0.102[a] ± 0.05	0.060[a] ± 0.02	1.70	0.162[a] ± 0.06	0.051[ab] ± 0.01	8.594[a] ± 3.12

Synthesis of high amount anthocyanins during heavy metal stress suggests that they have a critical role in detoxification mechanism of metal toxicities (Hale et al., 2001; Neill et al., 2002). It has been suggested that, anthocyanins especially have more critical performance in highly stressed conditions in compare to carotenoids. Researches revealed that carotenoids are able to protect plants against mild metal stress conditions (Dietz et al., 1999). Shahnawaz et al. (2017) treated barley with different concentrations of Al (2 mM, 4 mM and 6 mM) for 6 days. According to their results carotenoids decreased, but anthocyanins increased dose dependently. Similarly, Sevugaperumal et al. (2012) high anthocyanin content during Al toxicity in *Vigna radiata*.

Plants are face to different types of environmental stresses. Under stress conditions reactive oxygen species (ROS) being very reactive and toxic free radicals (O⁻, ·OH, HO·, $HO_2^·$ and ^1O) over-accumulate. ROS cause alterations in cellular functions unfavorably inducing peroxidation of membrane lipids, damage of protein and DNA, oxidation of carbohydrate, breakdown of photosynthetic pigments, distribution of antioxidant enzyme balance leading to programmed cell death (Gill and Tuteja, 2010; Woo et al., 2013; Nath and Lu, 2015). The antioxidant defense system consisted of several enzymatic (superoxide dismutase, catalase, ascorbate peroxidase, glutathione peroxidase, guaiacol peroxidase glutathione-S-transferase, etc.) and non-enzymatic antioxidants (ascorbic acid, glutathione, phenolic compounds, anthocyanins, alkaloids, carotenoids, etc.) providing balance between ROS production and scavenging (Gill and Tuteja, 2010). The disturbance in the balance between the production of ROS and antioxidant defenses culminates in oxidative stress. As an abiotic stress factor Al toxicity enhances the production and accumulation of ROS which trigger oxidative stress in cell (Sharma and Dubey, 2007; Gupta et al., 2013).

To assess the oxidative stress after Al application, superoxide dismutase (SOD) activity evaluated. It has been known that SOD which localize in cytosol and/or mitochondria catalyzes the reduction of superoxide anions ($O_2^{·-}$) into hydrogen peroxide (H_2O_2). $O_2^{·-}$ is an unstable signaling molecule and reacts with double-bounded biomolecules, Fe–S clusters of proteins and NO• (Domej et al., 2014; Das and Roychoudhury,

2014). According to our results SOD activity increased dose dependently. It was 8.2% in 50 μM, 10.2% in 100 μM, 22.5 in 150 μM and 42.9% in 200 μM. The highest increase was in 200 μM (Figure 2).

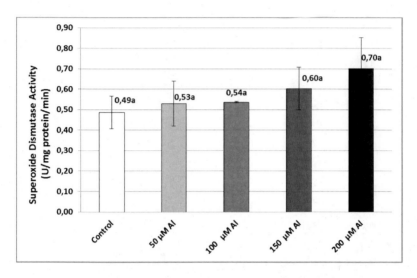

Figure 2. SOD activity in sunflower leaves after Al exposure. Different letters concern significance of independent samples at $P < 0.05$ level. ± SD: standard deviation.

Although $^1O_2$, $O_2^{\cdot-}$, and ·OH are almost instable and immobile because of their electrical charge, H_2O_2 is more stable and can cross biological membranes with a quite distance from its production site. H_2O_2 oxidizes proteins, reacts with $^1O_2$, and forms HO•. During stress conditions CAT catalyzes the breakdown of H_2O_2 to H_2O and O_2 (Domej et al., 2014; Das and Roychoudhury, 2014).

Based on our results, CAT activity increased in compare to control. It was increased by 1 fold in 50 μM, 1.5 fold in 100 μM, 1.5 fold in 150 μM and 1 fold in 200 μM. Even if CAT activity increased in compare to control at 200 μM, the highest enhancement observed at 150 μM (Figure 3).

Peroxidases (POD) are the most common scavenging enzymes to various types oxidative stress factors. They catalyze degradation of H_2O_2 to obtain highly oxidizing intermediates which oxidize various organic and inorganic substrates (Zaefyzadeh et al. 2009; Chen et al. 2010). However, our results revealed that POD activity reduced in compare to control. The reduction was 40% in 50 μM, 28% in 100 μM, 20% in 150 μM and 40% in 200 μM. The most reduction was observed in 50 and 200 μM Al treatment (Figure 4).

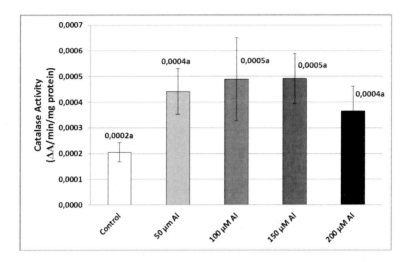

Figure 3. CAT activity in sunflower leaves after Al exposure. Different letters concern significance of independent samples at P < 0.05 level. ± SD: standard deviation.

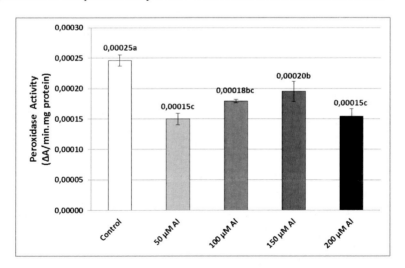

Figure 4. POD activity in sunflower leaves after Al exposure. Different letters concern significance of independent samples at P < 0.05 level. ± SD: standard deviation.

Previous results concerning Al toxicity, reveals difference between antioxidant enzyme activities based on concentration, duration and plant species (Panda et al., 2003; Boscolo et al., 2003; Sharma and Dubey, 2007; Bhoomika et al., 2013). Bhoomika et al. (2013) compared two rice (*Oryza sativa*) varieties differing in Al-tolerance. The researchers used 1 mM Al as moderate stress and 5 mM as higher stress for 5 days. According to their results activity of antioxidant enzymes such as SOD, CAT and POD increased in Al-resistant (cv Hur-105) variety. However, lower activity of antioxidant enzymes was observed in Al-tolerant (cv Vandana) variety. Besides, Sharma and Dubey (2007) treated rice (cv Pant-12) with 80 and 160 µM Al for 20 days. The results revealed that SOD and POD activities increased, but CAT activity decreased. Panda et al. (2003) treated *Vigna radiate* with different concentrations of Al (0.001 - 0.01 - 0.1 and 1 mM) for 48 h. Based on their results, SOD and POD activities increased, but CAT activities reduced. Similar results were also revealed by Boscolo et al. (2003) in two maize (*Zea mays*) varieties differing in Al-tolerance. SOD and POD activity increased in both resistant and tolerant varieties at 36 µM Al application after 48 h, but no increase in CAT activity observed.

Our results revealed that H_2O_2 degradation occurred due to activation of CAT rather than POD in sunflower leaves. Based on recent studies, while SOD activity increase in general, CAT and POD activity may differ during degradation of H_2O_2. This situation can be considered as the cellular response to Al may differ according to plant species/varieties, dose and exposure time.

Single cell gel electrophoresis (comet assay) is a quite sensitive and reliable assay resulting in shorter time than common cytogenetic methods used in to analyze the rate of possible genotoxic damage within treated cells. The results of % DNA tail, olive tail moment and tail intensity are the most informative data to detect the DNA damage (Lin et al., 2007; Aydin et al., 2018).

The tail of the comet origins from the migrating relaxed DNA loops from the core in the course of electrophoresis. The distance of DNA loops evaluates the length of the comet tail and it is used to determine the rate of DNA damage. The product of the tail length and the fraction of total DNA in the comet tail are characterized as tail moment. Tail moment includes the smallest detectable comet tail length and intensity of DNA tail. Besides, the tail intensity presents the quantity of DNA breaks and it may be considered as a better predictor of DNA damage than tail length (Collins, 2004; Møller, 2006; Kovačević et al., 2007).

According to our results % DNA tail was 68.63% in positive control group. In experimental groups it enhanced by 60.78% in 50 µM Al, %72.55 in 100 µM Al, %58.82 in 150 µM Al and %66.67 in 200 µM Al with respect to negative control. Although the results represented fluctuations, the DNA tail rate increased in compare to control. The increase in olive tail moment was 15.6 fold in positive control. It was also increased by 9.5-fold in 50 µM Al, 12.5-fold in100 µM Al, 11.75-fold in150 µM Al and 11.25-fold in 200 µM Al with respect to negative control. The olive tail moment reduced slightly after 100µM Al exposure. The tail intensity increased by 15-fold in positive control. The increment was 7.7-fold in 50 µM Al, 11.6-fold in 100 µM Al, 10.3-fold in 150 µM and 9.2-fold in 200 µM Al with respect to negative control. Similar to DNA tail and olive tail moment the tail intensity regressed after 100 µM Al (Table 2, Figure 5, 6).

Table 2. % DNA Tail, olive tail moment and tail intensity results of sunflower leaves after Al exposure. Different letters concern significance of independent samples at P < 0.05 level. ± SD: standard deviation

	% DNA Tail	Olive Tail Moment	Tail Intensity
Negative Control	$51^a \pm 19$	$4^a \pm 3$	$215^a \pm 130$
Positive Control	$86^{cd} \pm 9$	$63^d \pm 37$	$3228^e \pm 2702$
50 µM Al	$82^{bc} \pm 12$	$38^b \pm 28$	$1648^b \pm 1399$
100 µM Al	$88^d \pm 8$	$50^c \pm 32$	$2504^c \pm 2080$
150 µM Al	$81^b \pm 14$	$47^{bc} \pm 35$	$2205^{bc} \pm 2056$
200 µM Al	$85^{bcd} \pm 10$	$45^e \pm 26$	$1978^{bc} \pm 1555$

Multiple studies concerned Al-induced nuclear disruption, mitotic regression and chromosomal abnormalities (Campos and Vicini, 2003; Pan et al., 2004; Zheng and Yang, 2005; Vardar et al., 2011). Whereas the comet assay is one of the simple, fast, sensitive and reliable method estimating the DNA damage, only a few data are available on Al toxicity and comet assay in plants (Vardar et al., 2015; Aydin et al., 2018). The acute breaks in double strand DNA even at a single cell level can be detected by comet assay revealing the toxicity rate between environmental toxicants (Collins, 2004; Kovačević et al., 2007; Nikolova et al., 2013). The amount of breaks (tail intensity) in double stranded DNA also refers to programmed cell death.

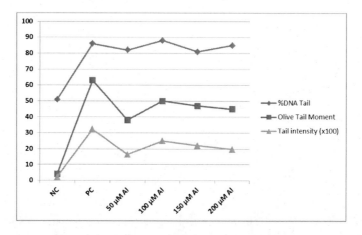

Figure 5. Comparison of % DNA tail, olive tail moment and tail intensity results of sunflower leaves after Al exposure. NC: negative control, PC: positive control.

Al-induced DNA breaks were reported in *Allium cepa* roots (Achary et al., 2008; Achary and Panda; 2009) by comet assay exposed to different concentrations of Al (1, 10, 50, 100, 200, 400 and 800µM) for short durations. According to their results dose dependent increase in tail length and olive tail moment were evaluated. Similar results were also revealed by Achary et al. (2012) in *Hordeum vulgare* roots. Vardar et al. (2015) compared the time dependent response of wheat, rye and triticale to 100 µM Al toxicity. According to their results DNA damage were detected within 15 min after Al treatment in three Gramineae species which can be considered as early signal of cell death. Consistent with the previous studies, Al induced DNA damage in sunflower leaves after long-term treatment.

In conclusion; different concentrations of Al (50, 100, 150 and 200 µM) induced alterations in sunflower leaves. According to our results, it can be concluded that concentrations up to 150 µM constituted moderate stress and sunflower leaves showed an adaptive response. However 200 µM is higher stress conditions that blocking its balance.

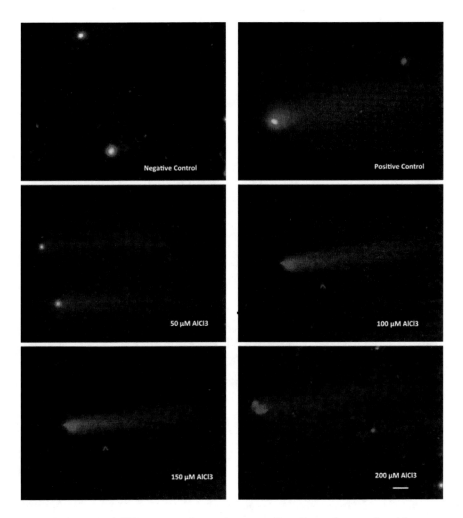

Figure 6. Comets of different experimental groups of sunflower leaves after Al exposure. Bar: 5 mm.

ACKNOWLEDGMENT

This research is performed by the support of Research Foundation of Marmara University (BAPKO no. FEN-A-081117-0624).

REFERENCES

Abate, E., Hussien, S., Laing, M. & Mengistu, F. (2013). Aluminum toxicity tolerance in cereals: Mechanisms, genetic control and breeding methods. *African Journal of Agricultural Research*, 8, 711-722.

Achary, V. M. M., Jena S., Panda K. K. & Panda, B. B. (2008). Aluminium induced oxidative stress and DNA damage in root cells of *Allium cepa* L. *Ecotoxicology and Environmental Safety*, 70, 300-310.

Achary, V. M. M. & Panda, B. B. (2009). Aluminium-induced DNA damage and adaptive response to genotoxic stress in plant cells are mediated through reactive oxygen intermediates. *Mutagenesis*, 25, 201-209.

Achary, V. M. M., Patnaik, A. R. & Panda, B. B. (2012). Oxidative biomarkers in leaf tissue of barley seedlings in response to aluminium stress. *Ecotoxicology and Environmental Safety*, 75, 16-26.

Arnon, D. I. (1949). Copper enzymes in isolated chloroplasts, polyphenoxidase in *Beta vulgaris*. *Plant Physiology*, 24, 1-15.

Aydin, Y., Uncuoğlu, A. A. & Vardar, F. (2018). Single cell gel electrophoresis for determination of plant responses to genotoxic stress. In: Yuksel, B; Karagül, MS. *Advances in Health and Natural Sciences* Nova Science Publishers, USA, Vol *12*.

Baek, S. A., Han, T., Ahn, S. K., Kang, H., Cho, M. R., Lee, S. C. & Im, K. H. (2012). Effects of heavy metals on plant growths and pigment contents in *Arabidopsis thaliana*. *Plant Pathology Journal*, 28, 446-452.

Bhoomika, K., Pyngrope, S. & Dubey, R. S. (2013). Differential responses of antioxidant enzymes to aluminum toxicity in two rice (*Oryza sativa* L.) cultivars with marked presence and elevated activity of Fe

SOD and enhanced activities of Mn SOD and catalase in aluminum tolerant cultivar. *Plant Growth and Regulation, 71*, 235-252.

Birecka, H., Briber, K. A. & Catalfamo, J. L. (1973). Comparative studies on tobacco pith and sweet potato root isoperoxidases in relation to injury, indoleacetic acid, and ethylene effects. *Plant Physiology, 52*, 43-49.

Boscolo, P. R. S., Menossi, M. & Jorge, R. A. (2003). Aluminum-induced oxidative stress in maize. *Phytochemistry, 62*, 181-189.

Cakmak, I. & Marschner, H. (1992). Magnesium deficiency and high light intensity enhance activities of superoxide dismutase, ascorbate peroxidase, and glutathione reductase in bean leaves. *Plant Physiology, 98*, 1222-1227.

Campos, J. M. S. & Vicini, L. F. (2003). Cytotoxicity of aluminum on meristematic cells of *Zea mays* and *Allium cepa*. *Caryologia, 56*, 65-73.

Cho, Y. W., Park, E. H. & Lim, C. J. (2000). Glutathione S-transferase activities of S-type and L-type thiol transferase from *Arabidopsis thaliana*. *Journal of Biochemistry and Molecular Biology, 33*, 179-183.

Chen, Q., Zhang, M. & Shen, S. (2010). Effect of salt on malondialdehyde and antioxidant enzymes in seedling roots of *Jerusalem artichoke* (*Helianthus tuberosus* L.). *Acta Physiology Plantarum, 33*, 273-278.

Ciamporova, M. (2002). Morphological and structural responses of plant roots to aluminum at organ, tissue and cellular levels. *Biologia Plantarum, 45*, 161-171.

Collins, A. R. (2004). The comet assay for DNA damage and repair. *Molecular Biotechnology, 26*, 249-261.

Das, K. & Roychoudhury, A. (2014). Reactive oxygen species (ROS) and response of antioxidants as ROS-scavengers during environmental stress in plants. *Frontiers in Environmental Science, 2*, 53.

Delhaize, E., Ryan, P. R., Hebb, D. M., Yamamoto, Y., Sasaki, T. & Matsumoto, H. (2004). Engineering high-level aluminum tolerance in barley with the ALMT1 gene. *PNAS, 101*, 15249-54.

Dietz, K. J., Baier, M. & Kramer, U. (1999). Free radicals and reactive oxygen species as mediator of heavy metal toxicity in plants. In: Prasad, MNV; Hagemeyer, J. *Heavy metal stress in plants: From molecules to ecosystem*, Springer – Verlag, Berlin, Germany, pp. 73-79.

Domej, W., Oettl, K. & Renner, W. (2014). Oxidative stress and free radicals in COPD – implications and relevance for treatment. *International Journal of Chronic Obstructive Pulmonary Disease*, 9, 1207-1224.

Frantzios, G., Galatis, B. & Apostolakos, P. (2000). Aluminum effects on microtubule organization in dividing root tip cells of *Triticum turgidum*. I. Mitotic cells. *New Phytology*, 145, 211-224.

Frantzios, G., Galatis, B. & Apostolakos, P. (2001). Aluminum effects on microtubule organization in dividing root tip cells of *Triticum turgidum*. II. Cytokinetic cells. *Journal of Plant Research*, 114, 157-170.

Gill, S. S. & Tuteja, N. (2010). Reactive oxygen species and antioxidant machinery in abiotic stres tolerance in crop plants. *Plant Physiology and Biochemistry*, 48, 909-930.

Gumze, A., Vinkovic, T., Petrovic, S., Eded, A. & Rengel, Z. (2007). Aluminium toxicity in maize hybrids during germination. *Cereal Research Communications, Alps-Adria Scientific Workshop*, 35, 421-424.

Gupta, D. K., Palma, J. M. & Corpas, F. J. (2013). *Reactive Oxygen species and oxidative damage in plants under stress*. Springer.

Gupta, P. & Bhatnagar, A. K. (2015). Spatial distribution of arsenic in different leaf tissues and its effect on structure and development of stomata and trichomes in mung bean, *Vigna radiata* (L.) Wilczek. *Environmental and Experimental Botany*, 109, 12-22.

Haider, S. I., Kang, W., Ghulam, J. & Guo-ping, Z. (2007). Interactions of cadmium and aluminum toxicity in their effect on growth and physiological parameters in soybean. *Journal of Zhejiang University Science B*, 8, 181-188.

Hale, K. L., McGrath, S. P., Lombi, E., Stack, S. M., Terry, N., Pickering, I. J., George, G. N. & Pilon-Smits, E. A. H. (2001). Molybdenum sequestration in *Brassica* species. A role for anthocyanins? *Plant Physiology, 126*, 1391-1402.

Hoagland, D. R. & Arnon, D. I. (1950). The water culture method for growing plants without soil. *Journal Circular*, Berkeley, California, Vol *347*.

Jamal, S. N., Iqbal, M. Z. & Athar, M. (2006). Phytotoxic effect of aluminium and chromium on the germination and early growth of wheat (*Triticum aestivum*) varieties Anmol and Kiran. *International Journal of Environmental Science and Technology, 3*, 411-416.

Jiang, H. X., Chen, L. S., Zheng, J. G., Han, S., Tang, N. & Smith, B. R. (2008). Aluminium induced effect on photosystem Π photochemistry in citrus leave assessed by the chlorophyll a fluorescence transient. *Tree Physiology, 28*, 1868-1871.

Jalili, M., Mokhtari, M., Eslami, H., Abbasi, F., Ghanbari, R. & Ebrahimi A. A. (2019). Toxicity evaluation and management of co-composting pistachio wastes combined with cattle manure and municipal sewage sludge. *Ecotoxicology and Environmental Safety, 171*, 798-804.

Joutchev, G., Menke, M. & Schubert, I. (2001). The comet assay detects adaptation to MNU-induced DNA damage in barley. *Mutation Research, 493*, 95-100.

Karimaei, M. & Poozesh, V. (2016). Effects of aluminum toxicity on plant height, total chlorophyll (Chl a+b), potassium and calcium contents in spinach (*Spinacia oleracea* L.) *International Journal of Farming and Allied Sciences, 5*, 76-82.

Kasim, W. A. (2006). Changes induced by copper and cadmium stress in the anatomy and grain yield of *Sorghum bicolor* (L.) Moench. *International Journal of Agriculture and Biology, 8*, 123-128.

Kasim, W. A. (2007). Physiological consequences of structural and ultrastructural changes induced by Zn stress in *Phaseolus vulgaris*. I. Growth and photosynthetic apparatus. *International Journal of Botany, 3*, 15-22.

Kinraide, T. B. & Parker, D. R. (1987). Cation amelioration of aluminum toxicity in wheat. *Plant Physiology*, *83*, 546-551.

Kovačević, G., Želježic, D., Horvatin, K. & Kalafatić, M. (2007). Morphological features and comet assay of green and brown hydra treated with aluminum. *Symbiosis*, *44*, 145-152.

Krishna Raj, S., Dan, T. V. & Saxena, P. K. (2000). A fragrant solution to soil remediation. *International Journal of Phytoremediation*, *2*, 117-132.

Li, H., Yang, L. T., Qi, Y. P., Guo, P., Lu, Y. B. & Chen, L. S. (2016). Aluminum toxicity-induced alterations of leaf proteome in two citrus species differing in aluminum tolerance. *International Journal of Molecular Sciences*, *17*, 1180.

Lin, A., Zhang, X., Chen, M. & Cao, Q. (2007). Oxidative stress and DNA damages induced by cadmium accumulation. *Journal of Environmental Science*, *19*, 596-602.

Luo, Y., Liang, J., Zeng, G., Chen, M., Mo, D., Li, G. & Zhang, D. (2018). Seed germination test for toxicity evaluation of compost: Its roles, problems and prospects. *Waste Management*, *71*, 109-114.

Ma, J. F., Chen, Z. C. & Shen, R. F. (2014). Molecular mechanism of Al tolerance in Gramineous plants. *Plant and Soil*, *381*, 1-12.

MacFarlane, G. R. (2003). Chlorophyll a fluorescence as a potential biomarker of zinc stress in the grey mangrove, *Avicennia marina*. *Bulletin of Environmental Contamination and Toxicology*, *70*, 90-96.

Matsumoto, H. (2000). Cell biology of aluminum toxicity and tolerance in higher plants. *International Review of Cytology*, *200*, 1-47.

Møller, P. (2006). Assessment of reference values for DNA damage detected by the comet assay in human blood cell DNA. *Mutation Research*, *612*, 84-104.

Nath, K. & Lu, Y. (2015). A Paradigm of reactive oxygen species and programmed cell death in plants. *Journal of Cell Science and Therapy*, *6* (2).

Neill, S. O., Gould, K. S., Kilmartin, P. A., Mitchell, K. A. & Markham, K. R. (2002). Antioxidant activities of red versus green leaves in *Elatostema rugosum*. *Plant Cell Environment*, *25*, 539-547.

Nikolova, I., Georgieva, M., Stoilov, L., Katerova, Z. & Todorova, D. (2013). Optimization of neutral comet assay for studying DNA double-strand breaks in pea and wheat. *Journal of BioScience and Biotechnology*, *2*, 151-157.

Pan, J. W., Zheng, K., Ye, D., Yi, H. L., Jiang, Z. M., Jing, C. T., Pan, W. H. & Zhu, M. Y. (2004). Aluminum-induced ultraweak luminescence changes and sister chromatid exchanges in root tip cells of barley. *Plant Science*, *167*, 1391-1399.

Panda, S. K., Singha, L. B. & Khan, M. H. (2003). Does aluminium phytotoxicity induce oxidative stress in greengram (*Vigna radiata*)? *Bulgarian Journal of Plant Physiology*, *29*, 77-86.

Pereira, L. B., Tabaldi, L. A., Goncalves, J. F., Jucoski, G. O., Pauletto, M. M., Weis, S. N., Nicoloso, F. T., Borher, D., Rocha, J. B. T. & Schetinger, M. R. C. (2006). Effect of aluminium on amino levulinic acid dehydrotase (ALA- D) and the development of cucumber (*Cucumis sativus*). *Environmental and Experimental Botany*, *57*, 106-115.

Poschenrieder, C., Gunsé, B., Corrales, I. & Barceló, J. (2008). A glance into aluminium toxicity and resistance in plants. *Science of Total Environment*, *400*, 356-368.

Radić, S., Babić, M., Skobić, D., Roje, V. & Pevalek-Kozlina, B. (2010). Ecotoxicological effects of aluminum and zinc on growth and antioxidants in *Lemna minor* L. *Ecotoxicology and Environmental Safety*, *73*, 336-342.

Radoglou, K. M. & Jarvis, P. G. (1990). Effects of CO_2 enrichment on four poplar clones. II. Leaf surface properties. *Annals of Botany*, *65*, 627.

Rucińska-Sobkowiak, R. (2016). Water relations in plants subjected to heavy metal stresses. *Acta Physiology Plantarum*, *38*, 257.

Sagardoy, R., Vázquez, S., Florez-Sarasa, I. D., Albacete, A., RibaFs-Carbó, M., Flexas, J., Abada, J. & Morales, F. (2010). Stomatal and mesophyll conductances to CO_2 are the main limitations to photosynthesis in sugar beet (*Beta vulgaris*) plants grown with excess zinc. *New Phytologist*, *187*, 145-158.

Samad, R., Rashid, P. & Karmoker, J. L. (2017). Effects of aluminium toxicity on germination of seeds and its correlation with K⁺, Cl⁻ and Al³⁺ accumulation in radicle and plumule of *Oryza sativa* L. and *Cicer aeriatinum* L. *Bangladesh Journal of Botany*, *46*, 979-986.

Sevugaperumal, R., Selvaraj, K. & Ramasubramanian, V. (2012). Removal of aluminium by padina as bioadsorbant. *International Journal of Biological and Pharmaceutical Research*, *3*, 610-615.

Shahnawaz, M. D., Rajani, C. & Sanadhya, D. (2017). Impact of aluminum toxicity on physiological aspects of barley (*Hordeum vulgare* L.) cultivars and its alleviation through ascorbic acid and salicylic acid seed priming. *International Journal of Current Microbiology and Applied Science*, *6*, 875-891.

Sharma, P. & Dubey, R. S. (2007). Involvement of oxidative stress and role of antioxidative defense system in growing rice seedlings exposed to toxic concentrations of aluminum. *Plant Cell Reports*, *26*, 2027-2038.

Shaw, C. A. & Tomljenovic, L. (2013). Aluminum in the central nervous system (CNS): toxicity in humans and animals, vaccine adjuvants, and autoimmunity. *Immunologic Research*, *56*, 304-316.

Shen, L., Sun, P., Bonnell, V. C., Edwards, K. J., Hetherington, A. M., McAinsh, M. R. & Roberts, M. R. (2015). Measuring stress signaling responses of stomata in isolated epidermis of Graminaceous species. *Frontiers in Plant Science*, *6*, 533.

Silva, I., Smyth, T., Moxley, D., Carter, T., Allen, N. & Rufty, T. (2000). Aluminum accumulation at nuclei of cells in the root tip. Fluorescence detection using lumogallion and confocal laser scanning microscopy. *Plant Physiology*, *123*, 543-552.

Singh, N. B., Yadav, K. & Amist, N. (2011). Phytotoxic effect of aluminium on growth and metabolism of *Pisum sativum* L. *International Journal of Innovations in Biological and Chemical Sciences*, *2*, 10-21.

Singh, S., Tripathi, D. K., Singh, S., Sharma, S., Dubey, N. K., Chauhan, D. K. & Vaculík, M. (2017). Toxicity of aluminium on various levels of plant cells and organism: A review. *Environmental and Experimental Botany, 137*, 177-193.

Sinha, S., Bhatt, K., Pandey, K., Singh, S. & Saxena, R. (2003). Interactive metal accumulation and its toxic effects under repeated exposure in submerged plant *Najas indica*. Cham. *Bulletin Environmental Contamination Toxicology, 70*, 696-704.

Smirnov, O. E., Kosyan, A. M., Kosyk, O. I. & Taran, Y. N. (2014). Buckwheat stomatal traits under aluminium toxicity. *Modern Phytomorphology, 6*, 15-18.

Strzalka, K., Kostecka-Guga, A. & Latowski, D. (2003). Carotenoids and environmental stress in plants: significance of carotenoid-mediated modulation of membrane physical properties. *Russian Journal of Plant Physiology, 50*, 168-172.

Tohidi, Z., Baghizadeh, A. & Enteshari, S. (2015). The effects of aluminum and phosphorous on some of physiological characteristics of *Brassica napus*. *Journal of Stress Physiology and Biochemistry, 11*, 16-28.

Tripathi, A. K., Pareek, A. & Singla-Pareek, S. L. (2016). TUNEL assay to assess extent of DNA fragmentation and programmed cell death in root cells under various stress conditions. *Plant Physioloy, 7*.

Vardar, F., Arıcan, E. & Gözükırmızı, N. (2006). Effects of aluminum on *in vitro* root growth and seed germination of tobacco (*Nicotiana tabacum* L.). *Advances in Food Science, 28*, 85-88.

Vardar, F. & Ünal, M. (2007). Aluminum toxicity and resistance in higher plants. *Advances in Molecular Biology, 1*, 1-12.

Vardar, F., İsmailoğlu, İ., İnan, D. & Ünal, M. (2011). Determination of stress responses induced by aluminum in maize (*Zea mays*). *Acta Biology of Hungarica, 62*, 156-170.

Vardar, F., Akgül, N., Aytürk, Ö. & Aydın, Y. (2015). Assessment of aluminum induced genotoxicity with comet assay in wheat, rye and triticale roots. *Fresenius Environmental Bulletin, 37*, 3352-3358.

Vardar, F., Çabuk, E., Aytürk, Ö. & Aydın, Y. (2016). Determination of aluminum induced programmed cell death characterized by DNA fragmentation in Gramineae species. *Caryologia, 69*, 111-115.

Vardar, F., Yanık, F., Çetinbaş-Genç, A. & Kurtuluş, G. (2018). Aluminum-induced toxicity and programmed cell death in plants. In: Yuksel, B; Karagül, MS. *Advances in Health and Natural Sciences* Nova Science Publishers, USA, Vol. *10*, pp. 155-182.

Vitorello, V. A., Capaldi, F. R. C. & Stefanuto, V. A. (2005). Recent advances in aluminum toxicity and resistance in higher plants. *Brazilian Journal of Plant Physiology, 17*, 129-143.

Weitbrecht, K., Muller, K. & Leubner-Metzger, G. (2011). First off the mark: early seed germination. *Journal of Experimental Botany, 62*, 3289-3309.

Woo, H. R., Kim, H. J., Nam, H. G. & Lim, P. O. (2013). Plant leaf senescence and death - regulation by multiple layers of control and implications for aging in general. *Journal of Cell Science, 126*, 4823-4833.

Yamamoto, Y. (2018). Aluminum toxicity in plant cells: Mechanisms of cell death and inhibition of cell elongation. *Soil Science and Plant Nutrition*, DOI:10.1080/00380768.2018.1553484

Zhang, F. Q., Wang, Y. S., Lou, Z. P. & Dong, J. D. (2007). Effect of heavy metal stress on antioxidative enzymes and lipid peroxidation in leaves and roots of two mangrove plant seedlings (*Kandelia candel* and *Bruguiera gymnorrhiza*). *Chemosphere, 67*, 44-50.

Zaefyzadeh, M., Quliyev, R. A., Babayeva, S. M. & Abbasov, M. A. (2009). The effect of the interaction between genotypes and drought stress on the superoxide dismutase and chlorophyll content in durum wheat landraces. *Turkish Journal of Biology, 33*, 1-7.

Zheng, S. J. & Yang, J. L. (2005). Target sites of aluminum phytotoxicity. *Biology Plantarum, 49*, 321-331.

Chapter 8

DETERMINATION OF GENOTOXIC EFFECTS OF ORGANOCHLORINE PESTICIDES IN WHEAT (*TRITICUM AESTIVUM* L.) BY COMET ASSAY

Melek Adiloğlu Öztürk and Yıldız Aydın[*]
Marmara University, Faculty of Science and Letters,
Department of Biology, Göztepe Campus, Istanbul, Turkey

ABSTRACT

Due to the detection of pesticides in underground and surface waters in many agricultural regions of the world, the environmental dimension of pesticide use is an important issue discussed today. Continuous use of pesticides in agriculture causes genetic resistance to pests. Pesticides threaten the ecosystem by altering the structure and species distribution of the ecosystem and disrupting the normal balance between food chains. Therefore pesticides need to be monitored under strict supervision.

In this study, our aim was to determine the genotoxic effects of endosulfan (ES) pesticide at different time and concentrations in wheat (*Triticum aestivum* L.). Leaf samples were taken from two-weeks old wheat seedlings of the negative control (Hoagland solution), positive

---
[*] Corresponding Author's Email: ayildiz@marmara.edu.tr.

control (Hoagland solution containing 0.1% H2O2) and different concentrations (1 g/L, 2 g/L and 4 g/L) of ES treated groups at different time intervals (6h, 12h and 18h) and examined by comet technique (single cell gel electrophoresis).

DNA damage levels increased significantly after the ES application in wheat seedlings time and dose dependently evaluated by comet analysis. In wheat seedlings, the highest DNA damage (% Tail DNA:50, Olive tail moment: 0.34) were determined at 12h of 2 g/L ES concentration ($p < 0.001$).

The obtained results of this study demonstrated that the ES is a genotoxic agent causing DNA breaks in wheat. Also, the results related with decrease in the level of DNA damage obtained from this study will contribute to the determination of the spraying and harvesting time as well as the determination of the appropriate concentrations using this pesticide.

Keywords: Endosulfan (ES), comet assay, *Triticum aestivum* L.

INTRODUCTION

The production of crop uses a wide variety of synthetically produced chemicals including insecticides, fungicides, herbicides and other pesticides to control plant diseases. Although most pesticide use is directed to the control of pests on above-ground plant parts, a large proportion of the pesticide reaches the soil. Pesticides can contaminate soil, water, turf, and other vegetation. In addition to killing insects or weeds, pesticides can be toxic to a host of other organisms including birds, fish, beneficial insects, and non-target plants. Plants can also suffer indirect consequences of pesticide applications when harm is done to soil microorganisms and beneficial insects.

In particular, broad-spectrum pesticides do not only harm pests, but also organisms that cause disease and off-target organisms. Due to the detection of pesticides in ground and surface waters in many agricultural regions of the world, the environmental dimension of pesticide use is an important issue discussed today.

Chemical control using pesticides is one of the most used methods in agricultural control. After unconsciously applied pesticides, the pesticides remaining on the soil surface can reach to the groundwater and other water resources by surface flow with rainwater or by washing downwards in the soil. Especially when used above the recommended doses, pesticide contaminates non-target organisms, causing genotoxic effects on organisms and adversely affecting the entire food chain.

Agronomic crops as well as other plants are deliberately exposed to pesticides and other chemicals applied in modern agriculture. These pollutants are hazardous for the plants. Wheat (*Triticum aestivum* L.) is one of the most important cereal crops in the world for both human food and animal feed. Wheat has become an important strategic product both economically and politically because of its wide adaptability, its ability to be machined, cheap, ease of production, transportation, storage and processing, and high nutritional value (Mut et al., 2007).

The most important biotic stress factors in wheat are disease agents of fungal origin such as *Tilletia foetida* (Sliding), *Puccinia recondita* (Pas), *Ustilago hordei*, *Septoria tritici*, *Rhizoctonia*, *Fusarium* and insects such as *Eurygaster*, *Aelia*, *Zabrus*, *Pachytychius hordei*. Chemical control using pesticides is most preferred in the fight against these pests. Because chemical control has high efficiency, it gives fast results and is economical when used consciously and controlled. These advantages are one of the main reasons why chemical control is still one of the most effective applications in plant protection.

Diseases such as rust diseases, randomness, root collar rot and pests such as insects are among the main reasons that negatively affect the production and yield of wheat which is the most cultivated grain in our country. In particular, the crop hump beetle (*Zabrus* spp.) Attracts crop leaves into the soil in the autumn, and in the spring, leaves and shoots cause significant loss of crops. Endosulfan (ES), an organochlorinated pesticide, is widely used in the chemical control of this parasite in wheat agriculture.

ES is an insecticide in the group of cyclodynes of broad-spectrum organochlorinated pesticides commonly used all over the world to protect pests from many crops including cereals, tea, coffee, cotton, fruit, oilseeds and vegetables (Herrmann, 2002; Aprea et al., 2002).

Organochlorine pesticides are considered as a potential hazard for both human health and the environment due to their wide range of toxic side effects as well as their benefits (Golfinopoulos et al., 2003). Most organochlorinated pesticides are known as persistent organic pollutants (Hilber et al., 2008).

Organochlorine pesticides dissolve very little in water and dissolve in organic solvents, mineral, vegetable and animal oils. Because of these properties, organochlorine pesticides remain in the environment for a long time and accumulate in the fat tissues of living things (Tuncer et al., 2000). This accumulation can sometimes exceed the amount that can poison humans and animals in a single dose acutely.

Organochlorinated pesticides are known to be resistant to biodegradation and are adsorbed by plankton, algae, invertebrates, plants and fish along the food chain.

ES, an organochlorinated pesticide, is an important environmental problem because of its long-term persistence in the environment and toxic to many non-target organisms. ES consists of a mixture of alpha and beta endosulfan, two stereoisomers in a 7: 3 ratio (Herrmann, 2002). The half-life of ES in water is from 1 month to 6 months under anaerobic conditions and the half-life in soil can range from 44.5 months to 6 years under aerobic conditions (ATSDR, Agency of Toxic Substances and Disease Registry, 2000; GFEA-U, German Federal Environment Agency, 2007).

As with many other pesticides, the distribution and degradation of endosulfan is influenced by the environmental conditions in which it is found. ES is not directly photolyzed but can be converted to its metabolites by chemical hydrolysis under alkaline conditions such as sea water. ES is exposed to abiotic and biotic transformation processes in the environment. ES is oxidized to form endosulfan sulfate and endosulfan diol in plants and soil. The hydrolysis of endosulfan diol, particularly under alkaline conditions or microbially, occurs chemically, while the formation of endosulfan sulfate is usually mediated by microorganisms. ES changes the permeability of the membranes of the cells in the roots of the plant and consequently inhibits the root growth by clinging to the root radical. It causes burns at the edges and ends of the leaves and stunts of the shoots (IPCS, International Program on Chemical Safety, 1984).

Continuous use of pesticides in agriculture leads to genetic resistance in harmful organism. Pesticides have threatened the ecosystem by altering the structure and distribution of the ecosystem and disrupting the normal balance between the food chains and therefore need to be monitored closely.

In nature, living things are confronted with many biotic and abiotic stress factors throughout their lives. Various chemicals, especially pesticides, are preferred in the fight against these stress factors due to their ease of use and short-term effects. Due to the increasing use of this pesticide, different toxic effects that can be detected physiologically, biochemically, morphologically and molecularly can occur on both target organisms and different non-target organisms (Gichner et al., 2009). Determination of DNA damage and the degree of damage, if any, is important for demonstrating genotoxic effects. Genotoxicity assays are designed to detect compounds that induce directly or indirectly damage the genetic material by different mechanisms, being a fundamental requirement for the assessment mutagenicity toxicological characterization of a chemical (Repetto et al., 2009).

In recent years, a new molecular test system called "comet technique" or "uniform cell gel electrophoresis has been developed more and more in the genotoxic effects. Comet technique has applications in many areas such as aging, genetic toxicology and molecular epidemiology (Martin et al., 1993).

The Comet test, where direct DNA damage is measured. Single cell gel electrophoresis (SCGE) or so-called "Comet assay"; is a more sensitive and reliable method that results in shorter time than conventional cytogenetic methods used in measuring genotoxic risk ratios (Gichner et al., 2003; Lin et al., 2007). The Comet method has also the advantages such as allowing direct measurement at the DNA level and detecting genetic damage in a short time and directly. Use of this technique also extends the utility of plants in basic and applied studies in environmental mutagenesis. Versions such as N/N (Neutral Dissolution/Neutral

Electrophoresis), A/N (Alkaline Dissolution/Neutral Electrophoresis) and A/A (Alkaline Dissolution/Alkaline Electrophoresis, pH 13) developed for different pHs are also used in the Comet assays (Gichner and Plewa, 1998; Lin et al., 2007). Comet is an effective tool that can be applied easily to fungi, algae, plants, sea creatures, insects, vertebrate animals and humans, and to give information on environmental health (Gichner and Plewa 1998, Angelis et al., 2000; Menke et al., 2001; Dhawan et al., 2009). The approach consists of an analysis of electrophoretic migration of DNA from nucleoids obtained after cell lysis in a thin layer of agarose. The assay, which combines gelelectrophoresis and fluorescent microscopy, usually starts with the preparation of microscopic slides with cells embedded in a thin layer of agarose. At the next step the immobilized cells are lysed with detergents and high ionic strength to produce so-called nucleoids that consist of DNA attached to some residual nuclear proteins that are insensitive to the lysis treatment. The slides are then electrophoresed: under the electric field DNA migrates from the nucleoid to the anode and forms an electrophoretic track, which, after staining with a fluorescent dye, looks like a comet tail (Östling and Johanson, 1984; Singh et al., 1988; Olive, 2002; Collins, 2004; Collins et al., 2008; Afanasieva and Sivolob, 2018).

The current use of comet assays appears to be oriented towards active monitoring and the exploration of genotoxic resistance, mainly involving DNA repair mechanisms (Azqueta and Collins, 2013). The first reports on the use of comet assay in plants date from the1990's (e.g., Cerda etal., 1993; Koppen and Verschaeve, 1996; Navarrete et al.,1997; Koppen and Angelis, 1998). The localized presences of characteristic meristematic regions (e.g., the concentration of highly dividing cells in the root apex) and the fact that root is usually the organ directly in contact with contaminated soil and water, have also influenced the establishment of plant comet assays in ecotoxicological approaches. In this study, our aim was to determine the genotoxic effects of endosulfan (ES) pesticide at different time and concentrations in wheat (*Triticum aestivum* L.).

MATERIAL AND METHODS

Wheat seeds which were surface sterilized placed in petri dishes to germinate for 15 days in distilled water. Plants were exposed to four doses of ES (1.0, 2.0, and 4.0 g/L) in Hoagland solution (Hoagland and Arnon, 1950) during 6, 12 and 18 h of exposure. Determination of DNA damage was performed following comet assay (alkaline single cell gel electrophoresis) method described by Jouvtchev et al. (2001) with minor modifications. Treatment with 0.1% H_2O_2 for 24 hours was determined as a positive control. After the exposure the leaves harvested and placed in a 60-mm petri dish on ice. Using a new razor blade, each leaf was gently cut into pieces in 400 μL of PBS. Nuclei were collected in the buffer with micropipettes and used in the comet assay. The cell suspension was mixed with 100 μL of 0.65% LMA in PBS at 37°C and pipetted onto fully frosted slides pre-coated with a layer of 0.65% HMA. To unwind the DNA, slides were incubated for 20 min in electrophoretic buffer containing 1 mM Na_2EDTA and 300 mM NaOH, pH 13. Then electrophoresis was run at 15 V and 300 mA for 30 min at 4°C. After electrophoresis the slides were neutralized in cold neutralization buffer (0.4 M Tris, pH 7.5) three times at 5-min (Figure 6). Prior to examination, slides were stained with 75 μL ethidium bromide (10 μg/mL) for analysis. Approximately 50 cells per slide were randomly scored with each providing a median value of %DNA in tail, olive tail moment using an image analysis system (BAB BS200ProP) attached to a fluorescent microscope equipped with appropriate filters.

RESULTS AND DISCUSSION

The acute genotoxic effects of ES toxicity was evaluated for 1.0, 2.0, and 4.0 g/L treatments in wheat at 6, 12 and 18 h. To assess the degree of possible DNA fragmentation/damage within treated cells, the widely used single cell electrophoresis (comet assay) was performed. Based on being able to detect a significant change in treated cells, the DNA tail, olive tail moment and tail intensity appeared to be the most informative measure of DNA damage.

PH, temperature and electrical conductivity of the media were checked during the applications and no significant change was observed in all parameters since ES lost its effect by degrading under high alkaline conditions (Figure 1).

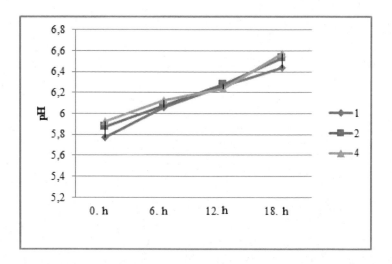

Figure 1. Changes in the pH of the medium during the application of ES at different concentrations and periods to two-week wheat seedlings.

Similarly, in our study carried out at constant temperature in order to prevent the degradation of ES due to temperature increase, the absence of a certain change in electrical conductivity is important in terms of showing that the ES does not suffer from degradation (Figure 2).

The positive, negative controls of wheat seedlings and the application groups (1.0, 2.0 and 4.0 g/L ES) at the 6th, 12th and 18th hours showed comet formations reflecting the DNA profile that had migrated off the nucleus due to genotoxic damage. When the comet images obtained were examined, a significant tail structure was observed when compared with the negative control (Figure 3 A) due to the genotoxic agent (H_2O_2) applied in the positive control group (Figure 3 B). The visual comet scores of all results in which the genotoxic damage caused by concentration and time can be observed visually in ES application groups at different concentrations are given in Figure 3.6 C-K. All images were analyzed with the help of BAB BsPro200 computer program and the results were evaluated with% DNA tail and olive tail moment parameters which provide a good correlation in genotoxicity studies.

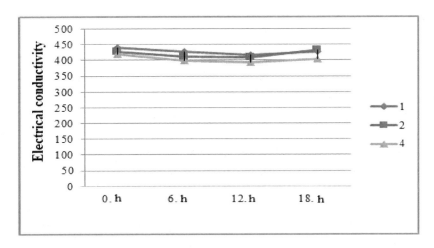

Figure 2. Changes in the electrical conductivity of the medium during the application of ES at different concentrations and periods to two-week wheat seedlings.

It was determined that DNA damage levels due to ES exposure and determined by comet technique increased significantly at all hours and concentrations compared to negative control group. Accordingly, the highest DNA damage level (% DNA tail: 50, Olive tail moment: 0.34) was determined in wheat samples treated with 2 g/L ES at 12 hours ($p < 0.001$). There was also a nonlinear decrease in the level of DNA damage determined in all groups depending on the application time. In a similar study, Akcha et al. (2008) observed the highest DNA damage at 1.0 µg/L ES when exposed to 1, 10 and 100 µg/L ES for 24 hours in a phytoplankton variety (Karenia mikimotoi) and a concentration-dependent increase reported that they do not. These results support a reduction in the level of DNA damage at all concentrations except 2 g/L ES at 12 hours due to increased concentration in this study. There was a significant decrease in the level of DNA damage (% DNA tail: 43, Olive tail moment: 0.26) in the

6th hour (% DNA tail: 26, Olive tail moment: 0.29) of 1 g/L ES application (Figures 4, 5).

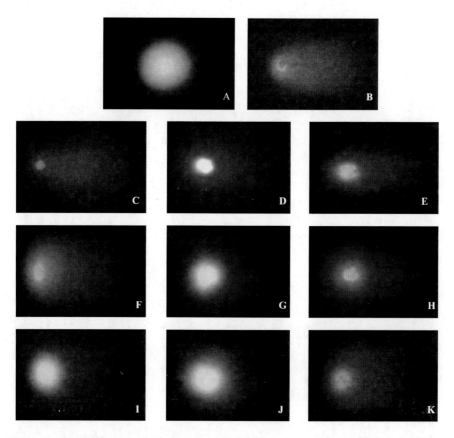

Figure 3. Detection of DNA damage exposure by ES exposure at different concentrations and times under fluorescence microscope in barley.**A**:Negative control,**B**: Pozitive control **C**: 1 g/L ES 6. h **D**: 1 g/L ES 12. h **E**: 1 g/L ES 18. h, **F**: 2 g/L ES 6. h, **G**: 2 g/L ES 12. h, **H**: 2 g/L ES 18. h, **I**: 4 g/L ES 6. h, **J**: 4 g/L ES 12. h, **K**: 4 g/L ES 18. h.

Liu et al. (2009) examined the genotoxic effects of different concentrations of ES (0.1,1.0 and 10.0 mg/L) on white alfalfa (*Trifolium repens* L.) and soil worm (*Elisenia foetida*) by using comet technique. reported genotoxic damage in all treatment groups and reported that the highest DNA damage was seen in the 10.0 mg/L application group. Similarly, in this study, 2 g/L ES concentration of the highest DNA damage level was determined at 12 hours (% DNA tail: 50, Olive tail moment: 0.34).

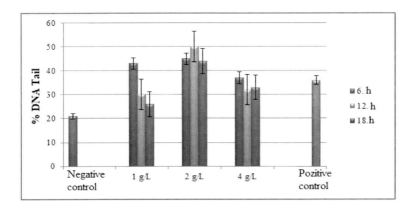

Figure 4. Percentage of DNA damage resulting from ES exposure in wheat plants at different concentrations and times (% DNA Tail).

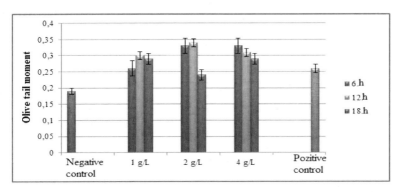

Figure 5. Percentage of DNA damage resulting from ES exposure in wheat plants at different concentrations and times (% DNA Tail moment).

Perez et al. (2013) exposed wetland high-structured *Bidens laevis* L. to ES pesticide at concentrations of 5, 10, 50 and 100 µg/L for 24 hours and examined the frequency of chromosome aberrations and found that genetic damage increased due to increased concentration they have.

Chlorinated compounds cause H_2O_2 production in plant tissues and this has been demonstrated by studies with chlorinated xenobiotics such as ES, Trichlorophenol, Atrazine (Menone et al., 2008; Michalowicz and Duda 2009). Liu et al. (2003) showed that triclorobenzene causes micronucleus and chromosomal aberrations due to oxidative stress in stem end cells in germination of pods (*Vicia faba*). Liu et al., (2004) in another study they have found that trichlorobenzene soybean (*Glycine max*) stem-end cells in the nucleus of the DNA chain breaks have been detected by the comet technique.

Jovtchev et al. (2001) treated the barley plant with MNU, an alkylating agent, examined DNA damage in leaf and root tip nuclei by comet test and observed a dose-dependent increase. They found that there was a significant decrease in the amount of DNA migration after the 6th hour in the root end nucleus and in the 18th hour in the leaf nucleus. In this study, the highest DNA damage was obtained in 1,0 g/L concentration of endosulfan at 6h. The lowest DNA damage was obtained in 1,0 g/L concentration of endosulfan at 18h. According to Tail length, at the 18th hours DNA damage was high in 2.0 g/L endosulfan concentration rather than the others. At 12th hours in both Tail length and Tail moment DNA damage was low in 4,0 g/L and high in 2,0 g/L doses of endosulfan.

Long-term genotoxicity studies can be an important approach to gain insight into the ability of organisms to repair DNA and other protective mechanisms. This study indicated that the alkaline Comet assay is a sensitive tool for evaluating the genotoxic effect of endosulfan in wheat leaves because significant effects of different endosulfan concentrations and duration of exposure were observed. Endosulfan exposure resulted in DNA damage for wheat nuclei.

In conclusion, the present study provided evidence that ES taken up by the root system of wheat generated DNA damage revealed by comet assay. The comet assay is sensitive enough to detect the effects of aluminum and could also provide one criterion for assessing the ES toxicity.

ACKNOWLEDGMENT

This research is performed by the support of Research Foundation of Marmara University (BAPKO no. FEN-C-YLP- 031210-0285).

REFERENCES

Afanasieva, K. and Sivolob, A. (2018). Physical principles and new applications of comet assay. *Biophysical Chemistry* 238: 1–7.

Angelis, K.J., McGuffie, M., Menke, M., Schubert, I. (2000). Adaptation to alkylation damage in DNA measured by the Comet assay. *Environmental and Molecular Mutagenesis*, 36: 146–50.

Aprea, C., Colosio, C., Mammone, T., Minoia, C., Maroni, M. (2002). Biological monitoring of pesticide exposure: A review of analytical methods. *Journal of Chromatography B*, 769, 191–219.

ATSDR. *Toxicological Profile for Endosulfan.* Agency of Toxic Substances and Disease Registry, Atlanta, USA (2000). http://www.atsdr.cdc.gov/toxprofiles/tp41.html.

Azqueta, A., Collins, A.R. (2013). The essential comet assay: a comprehensive guide to measuring DNA damage and repair. *Arch. Toxicol.* 87, 949e968.

Cerda H., von Hofsten B., Johanson K.J. (1993). Identification of irradiated foods by microgel electrophoresis of DNA from single cells. In: Leonardi M, Belliardo JJ, Raffi JJ (eds) *Recent advances of new methods of detection of irradiated foods.* Commission of the European Communities, Luxembourg, EUR 14315, pp 401–405

Collins, A.R., Oscoz, A.A., Brunborg, G., Gaivão, I., Giovannelli, L., Kruszewski, M., Smith, C.C., Štětina, R. (2008). *The comet assay: topical issues, Mutagenesis* 23: 143–151.

Collins, A.R. (2004). The comet assay for DNA damage and repair: principles, applications, and limitations, *Mol. Biotechnol.* 26: 249–261.

Dhawan, A., Bajpayee, M., Parmar, D. (2009). Comet assay: a reliable tool for the assessment of DNA damage in different models. *Cell Biol Toxicol.* 25:5-32.

Gichner, T. (2003). DNA damage induced by indirect and direct acting mutagens in catalase-deficient transgenic tobacco cellular and acellular Comet assays. *Mutation Research* 535: 187-193.

Gichner, T., Plewa, M.J. (1998). Induction of somatic DNA damage as measured by single cell gel electrophoresis and point mutation in leaves of tobacco plants. *Mutation Research*, 401: 143-152.

Gichner, T. (2009). Znider, I., Wagner, E.D., Plewa, M. "The use of higher plants in the comet assay", *The Comet Assay in Toxicology*, Edited by Alok Dhawan and Diana Anderson. *Royal Society of Chemistry*, p: 98- 119.

Golfinopoulos, K.S., Nikolaou, D.A., Kostopoulou, M.N., Xilourgidis, N.K., Vagi, M.C., Lekkas, D.T. (2003). Organochlorine pesticides in the surface waters of Northern Greece. *Chemosphere*, 50: 507–516.

Herrmann, M. (2002). *Preliminary risk profile of endosulfan Umweltbundesamt*; Berlin Germany.

Hilber, I., Mäder, P., Schulin, R., Wyss, G.S. (2008). Survey of organochlorine pesticides in horticultural soils and there grown Cucurbitaceae. *Chemosphere*, 73: 954–961.

Hoagland, D.R., Arnon, D.I. (1950). The water culture method for growing plants without soil. *Journal Circular,* Berkeley, California, Vol 347.

IPCS (1984). Endosulfan. World Health Organization. *International Programme on Chemical Safety (Environmental Health Criteria 40),* Geneva.

Jovtchev, G., Menke, M., Schubert, I. (2001). The comet assay detects adaptation to MNU-induced DNA damage in barley. *Mutation Research*, 493: 95-100.

Koppen, G., and Angelis, K.J. (1998). Repair of X-ray induced DNA damages measured by the Comet assay in roots of Vicia faba. *Environ. Mol. Mutagen.* 32, 281–285.

Koppen, G., and Verschaeve, L. (1996).The alkaline comet test on plant cells: a new genotoxicity test for DNA strand breaks in *Vicia faba* root cells. *Mutat.Res.* 360, 193–200. doi:10.1016/S0165-1161(96)90017-5

Lin, A., Zhang, X., Chen, M., Cao, Q. (2007). Oxidative stress and DNA damages induced by cadmium accumulation. *Journal of Environmental Science*, 19: 596-602.

Liu, W., Yang, Y.S., Li, P., Zhou, Q., Sun T. (2004). Root growth inhibition and induction of DNA damage in soybean (Glycine max) by chlorobenzenes in contaminated soil. *Chemosphere,* 57:101–106.

Liu, W., Zhou, Q.X., Li, P.J., Sun, T.H., Yang, Y.S., Xiong, X.Z., (2003). 1,2,4- Trichlorobenzene induction of chromosomal aberrations and cell division of root-tip cells in Vicia faba seedlings. *Bulletin of Environmental Contamination and Toxicology,* 71:689–697.

Liu, W., Zhu, LS., Wang, J., Wang, J.H., Xie, H., Song, Y. (2009). Assessment of the Genotoxicity of Endosulfan in Earthworm and White Clover Plants Using the Comet Assay. *Archive Environmental Contamination Toxicology,* 56:742–746.

Martin, V.J., Green, M.H.L., Schmezer, P., Zobel, B.L., De Meo, M.P., Collins, A. (1993). The Single Cell Gel Electrophoresis Assay (Comet Assay), A European Review. *Mutation Research,* 288: 47-63.

Menke, M.; Chen, I.P.; Angelis K.J.; Schubert, I. (2001). DNA damage and repair in Arabidopsis thaliana as measured by the Comet assay after treatment with different classes of genotoxins. *Mutat. Res.,* 493, 87.

Menone, M.L., Pesce, S.F., Díaz, M.P., Moreno, V.J., Wunderlin, D.A. (2008). Endosulfan induces oxidative stress and changes on detoxication enzymes in the aquatic macrophyte Myriophyllum quitense. *Phytochemistry,* 69, 1150-1157.

Michalowicz, J., Duda, W. (2009). The effects of 2,4,5-trichlorophenol on some antioxidative parameters and the activity of glutathione S-transferase in reed canary grass leaves (Phalaris arudinacea). *Polish Journal of Environmental Studies,* 18: 845–852.

Mut, Z., Aydın, N., Bayramoğlu, H.O., Özcan, H. (2007). Determination of Yield and Major Quality Characteristics of Some Common Wheat (Triticum aestivum L.) Genotypes, *Journal of Ondokuz Mayıs University Faculty of Agriculture,* 22: 193-201.

Navarrete, M.H., Carrera, P., Miguel, M., and Torre, C. (1997). A fast Comet assay variant for solid tissue cells. The assessment of DNA damages in higher plants. *Mutat. Res.* 389: 271–277.

Olive, P.L. (2002). The comet assay. An overview of techniques, *Methods Mol. Biol.* 203: 179–194.

Östling, O.K.J. (1984). Johanson, Microelectrophoretic study of radiation-induced DNA damages in individual mammalian cells, *Biochem. Biophys. Res. Commun.* 123: 291–298.

Perez, D., Lukaszewicz, G., Menone, M.L., Ame, M.V., Camadro, E.L., (2013). *Genetic and biochemical biomarkers in the macrophyte Bidens laevis L. exposed to a commercial formulation of endosulfan.* DOI: 10.1002/tox. 21836.

Repetto Jiménez, M. and Repetto Kuhn, G. (2009). *Toxicología Fundamental.* 4ª ed. Díaz de Santos, Madrid.

Singh, N.P., McCoy, M.T., Tice, R.R., Schneider, E.L. (1988). A simple technique for quantitation of low levels of DNA damage in individual cells, *Exp. Cell Res.* 175: 184–191.

Tuncer, C., Akça, İ., Saruhan, İ. (2000). Insecticides and their effects on the environment. *Agricultural Environment and Water Pollution Seminar.* 125-164, 26-28 September, Samsun.

In: A Closer Look at the Comet Assay
Editor: Keith H. Harmon

ISBN: 978-1-53611-028-9
© 2019 Nova Science Publishers, Inc.

Chapter 9

INVESTIGATION OF GENOTOXIC EFFECTS OF ORGANOPHOSPHORUS PESTICIDES IN BARLEY (*HORDEUM VULGARE* L.)

Tuba Akan and Yıldız Aydın[*]
Marmara University, Faculty of Science and Letters,
Department of Biology, Istanbul, Turkey

ABSTRACT

Chlorpyrifos [O O-diethyl-O-(3 5 6-trichloro-2-pyridyl)-Phosphorothioate (CPF)] is one of the most widely used organophosphate insecticides. Pesticides are increasingly used chemicals to enhance the yield and quality of agricultural products. With their increasing use, they pose a risk for both human health and ecosystem. While pesticides used against harmful effects of agricultural production affect the target organisms, they can affect the non-target organisms through wrong and especially excessive applications. Plant communities are often subjected to target as well as non-target exposure to pesticides. The Comet assay is used as a useful tool in assessing the potential of plants as source of information on the genotoxic impact of dangerous pollutants and sensitive sensors in ecosystems. This technique is based on the quantification of denatured DNA fragments migrating out of the

---
[*] Corresponding Author's Email: ayildiz@marmara.edu.tr.

cell nucleus during electrophoresis. This research was performed to determine the genotoxicity potential of Chlorpyrifos (CPF) at different concentrations and exposure times in barley by comet assay. The barley seedlings were treated with negative control (Hoagland's solution), positive control (Hoagland's solution with %0,1 H_2O_2) and different concentrations (5ml/L, 10ml/L, 20ml/L and 40 ml/L) of CPF for 6, 12 and 18h. In barley seedlings, the highest single stranded DNA breaks were determined for (%DNA tail: 55, Olive tail moment: 0.36) 6h of 5ml/L CPF dosage. During the exposure time, although the linear decrease in DNA damage has been observed in leaves of barley, the DNA damage was increased in 40ml/L CPF for 12 h (% DNA tail: 47, Olive tail moment: 0.29). Our research findings clearly indicated that CPF has genotoxic effects in barley. The findings may contribute to the determination of the appropriate doses for the use of ES, as well as the determination of spraying and harvesting time.

Keywords: Chlorpyrifos, comet assay, *Hordeum vulgare* L.

INTRODUCTION

Pesticides are actively used in pest control in order to obtain maximum yield and quality in the production of agricultural products. Because of the continuous application of pesticides for many years in agriculture leads to an increase in the resistance of pests to these drugs, the use of pesticides increases. As a result of intensive and unconscious use of pesticides, the pesticide itself or its conversion products may remain in the soil, water and air. In particular, pesticides remaining on the surface of the soil can reach the ground water and other water resources by surface run down with rainwater or by washing downwards in the soil. Exposure to pesticides occurs especially when used above the recommended doses.

It is undesirable for pesticides to degrade (biodegradations) before they reach target organisms, which can develop in a slow and long process and may develop as adsorption, transfer and decomposition (microbiological, chemical, photo-decomposition). It is also possible that the toxic transformation products resulting from this degradation sometimes have more harmful effects on organisms than the pesticide itself. Following the effect of pesticide on the target organism in the desired time period, the transformation of the pesticide into a non-toxic harmless form is important in terms of minimizing its environmental effects.

In order to minimize the risks posed by pesticides on ecosystems, it is important to use appropriate pesticides. An ideal pesticide should be harmful to non-target organisms when controlling the pest, turn into ecologically acceptable products, remain in the application area alone, have no potential to accumulate in the environment, and be reliable for users. However, no pesticide used today has all of these ideal qualities. As a result of the continuous change of ecosystem, the pesticides used become more harmful over time and have negative effects on the environment. Therefore, the use of some pesticides is restricted and even banned.

Chlorpyrifos (CPF) is a broad-spectrum organophosphorus insecticide used to combat insects and arthropods in agriculture (Hayes and Laws, 1990; Schardein and Scialli, 1999; Steenland et al., 2000). Organophosphorus pesticides, first synthesized by a group of German chemists in 1937, are generally less soluble in water and better soluble in oil and organic solvents. Organophosphorus compounds that lose their effects by hydrolysis in water can easily penetrate the fruits and leaves of plants because they dissolve well in oil (Daş, 2004).

CPF is widely used in grassy areas, for ornamental plants, farm structures and farm animal pests, in places such as homes, offices, etc. (extoxnet.orst.edu) in addition to agricultural areas. Because of this widespread use, CPF conversion products can remain in the environment, posing a danger to the entire ecosystem. CPF is toxic to some plants such as lettuce and the residue on these plants lasts for about 10-14 days on the plant surface. Studies have shown that this insecticide and its metabolites accumulate in some plants. Although field applications of algal areas are frequently followed, there is insufficient information about CPF toxicity on freshwater plants (Thomson, 1982).

Barley (*Hordeum vulgare* L.), which is one of the most important cultural plants in the prehistoric times, is still one of the economically important plants (Temel et al., 2008).

Barley has gained phenotypic properties that will facilitate agricultural production both through breeding studies and natural selection that it has been exposed to throughout history (Pourkheirandish and Komatsuda, 2007). In addition to these features, very cold and harsh climates, drought, alkali and salty soils to adapt to a high rate of stress and early maturation because of their characteristics such as early equatorial regions can be cultivated in a wide area (Schulte, 2009).

In nature, living things are confronted with many biotic and abiotic stress factors throughout their lives. Various chemicals, especially pesticides, are preferred in the fight against these stress factors due to their ease of use and short-term effects. Due to the increasing use of this pesticide, different toxic effects can be detected physiologically, biochemically, morphologically and molecularly occuring on both target organisms and different non-target organisms (Gichner et al., 2009). Determination of DNA damage and the degree of damage is important for demonstrating genotoxic effects.

Comet test can be applied very easily on fungi, algae, theoretically all to all plants, sea creatures, insects, vertebrates and humans. It is an important method used in monitoring of environmental health as well as providing information about the defense potentials and future health conditions of target organisms (Gichner and Plewa 1998; Angelis et al., 1999; Menke et al. 2001; Dhawan et al., 2009). Although almost all plants can be used for comet testing, potatoes (*Solanum tuberosum*), tobacco (*Nicotiana tabacum*), soybeans (*Vicia faba*), onions (*Allium cepa*) and *Arabidopsis thaliana* are both easy to obtain, breed and isolate their cells. (Gichner et al., 2006; Mancini et al., 2006; Lin et al., 2008). Gichner and Plewa (1998) found that after treating *Nicotiana tabacum* plants with 1-8 mM ethyl methane sulfonate (EMS), the plants exposed to the highest EMS concentration had the most DNA damage.

Menke et al. (2000) investigated that the N-methyl-N-nitroso Urea (MNU) was genotoxic effective agent in V. faba plants in comparison with three different comet technique protocols (neutral/neutral, alkaline/neutral, alkali/alkaline). Gichner et al. (2008) conducted that comet analysis under alkaline conditions to determine the increase in induced DNA damage in S.

tuberosum L. plant treated with EMS, Cd+2 and γ-rays, showed that the best dose-response curve was detected with the alkaline Comet technique.gichner (2003) found that tobacco roots from CAT1AS transgenic lines treated with precursor mutagen o-phenylenidamine (o-PDA) had higher levels of DNA damage than in SP1. Maluszynka and Juchimiuk (2005) used Comet analysis, sister chromatid exchange and FISH methods to detect and damage DNA by mutagen application to *Crepis capillaris*. Zhang et al. (2006) investigated the effects of *Glomus mossea*, a mycorrhizal fungus, on *Vicia faba* and the amount of toxicity triggered by heavy metals (Cu, Zn, Pb, Cd) contaminated with field soil and observed high DNA damage in mycorrhizal plants. Gichner et al. (2008), EMS, Cd² and γ-treated with *Solanum tuberosum* L. treated with the light-induced DNA damage to reveal the increase of DNA under Comet analysis and showed that the best dose-response curve was determined by the alkali comet technique.

DNA fragmentation being one of the important signatures of PCD originates from internucleosomal DNA cleavage by specific proteases and nucleases. DNA fragmentation can also be analyzed by TUNEL reaction, comet assay, laddering on the agarose gel and flow cytometry (Tripathi et al. 2016). Vardar et al. (2015; 2016) revealed DNA fragmentation using comet assay and agarose gel electrophoresis under Al stress in early hours in wheat, rye, barley, oat and triticale.

MATERIAL AND METHODS

Seeds of *Hordeum vulgare* L. were surface sterilized placed in petri dishes to germinate for 15 days (Figure 1). During growth, plants were routinely watered with Hoagland's nutrient solution. Many genotoxicity studies on plant based on root system, the other plant organs such as leaves are of secondary importance. Thus, we designed a study with different concentrations of CPF to reveal the responses of *Hordeum vulgare* leaves to CPF toxicity. Plants were exposed four doses of chlorpyrifos-ethyl (CPF) (0.5, 1, 2 and 4 ml/l) in Hoagland solution during 6, 12 and 18 h of exposure. Treatment containing only 0.1% H_2O_2 for 18 hours was the positive control. The Hoagland solution includes 5 mM $Ca(NO_3)_2$, 5 mM KNO_3, 2 mM $MgSO_4$, 1 mM KH_2PO_4, 30 μM Fe(III)-EDTA and standard Hoagland micronutrients (Hoagland and Arnon, 1950).

After exposure of barley to the different levels of pesticide, the leaves harvested and placed in a 60-mm petri dish on ice. Using a new razor blade, each leaf was gently cut into pieces in 400 μL of PBS. Nuclei were collected in the buffer with micropipettes and used in the comet assay. The cell suspension was mixed with 100 μL of 0.65% LMA in PBS at 37 °C and pipetted onto fully frosted slides precoated with a layer of 0.65% HMA. To unwind the DNA, slides were incubated for 20 min in electrophoretic buffer containing 1 mM Na_2EDTA and 300 mM NaOH, pH 13.Then electrophoresis was run at 15 V and 300 mA for 30 min at 4°C. After electrophoresis the slides were neutralized in cold neutralization buffer (0.4 M Tris, pH 7.5) three times at 5-min (Figure 6). Prior to examination, slides were stained with 75 μL ethidium bromide (10 μg/mL) for analysis. Approximately 50 cells per slide were randomly scored using an image analysis system (BAB BS200ProP)attached to a fluorescent microscope equipped with appropriate filters.

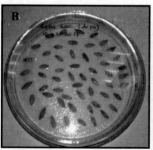

Figure 1. A. Surface sterilizationof seeds of *Hordeum vulgare* L. B-C:Seeds were placed in petri dishes to germinate for 15 days. During growth, plants were routinely watered with Hoagland's nutrient solution.

Determination of DNA damage was performed following comet assay (alkaline single cell gel electrophoresis) method described by Jouvtchev et al. (2001) with minor modifications. Hoagland without $AlCl_3$ was used for negative control and Hoagland with 0.1% H_2O_2 used for positive control group. The nuclei were isolated from control and Al treated sunflower leaves (0.4 g) by careful slicing with a razor blade in 0.4 ml Tris-HCl buffer (0.4 M, pH 7.5) on ice, in dark conditions. The homogenate filtered through sterile 20 micron filter. For each group, 100 µl of the nuclei suspension were mixed with 0.8 ml hot 0.65% low-melting point agarose (LMA) in a micro-centrifuge tube, and dropped onto slides pre-coated with 0.65% normal melting point agarose (HMA). Each drop was covered with a 24×60 mm cover slip and slides were stored at +4 °C for 10 min. After removal of the cover slips, the HMA-coated slides were placed submerged in alkaline buffer (0.3 M NaOH; 1 mM EDTA) for 15 min that facilitated nuclear DNA unwinding, followed by electrophoresis at 26 V/300 mA for 30 min, under cold and dark conditions. After electrophoresis, the slides were neutralized in 0.1 M PBS, three times for 5 min. Slides were stained by 0.2% ethidium bromide (EB). Samples were examined with a Leica DM LB2 fluorescence microscope, equipped with a 515-560 nm excitation filter and a 590 nm barrier filter for EB. The Comet module of the BAB BsPro200 image processing, and analyzing software were used for measurement of DNA content in the head and DNA tail of each comet. At least 2 slides from each independent experimental group were evaluated, with each providing a median value of %DNA in tail and olive tail moment of at least 50 comets. All experiments were carried out at least three times.

RESULTS AND DISCUSSION

In organisms exposed to genotoxic agents, some biological processes are adversely affected. Growth cessation, increases or decreases in protein and nucleic acid contents, inhibition of important enzymes, changes in pigment content, damage or mutations in DNA are some important biological responses that may occur in plants due to the effects of genotoxic agents. During the experiment, partial yellowing of the leaf tissues was observed morphologically due to different doses of CPF and 0.1% H_2O_2 (positive control) applications (Figure 2,3).

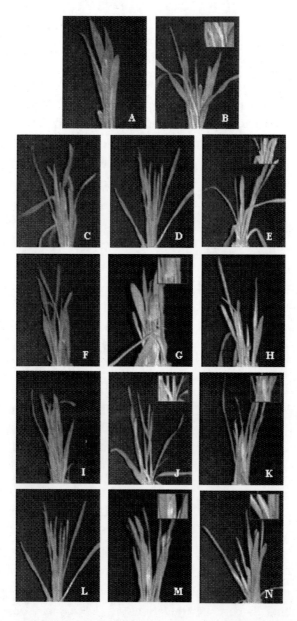

Figure 2. Yellowing of barley seedlings resulting from the application of CPF and 0.1% H_2O_2 as a positive control at different doses and times. **A:** Negative control, **B:** Positive control **C:** 5 ml/L CPF 6. h **D:** 5 ml/L CPF 12. h **E:** 5 ml/L CPF 18. h, **F:** 10 ml/L CPF 6. h, **G:** 10 ml/L CPF 12. h, **H:** 10 ml/L CPF 18. h, **I:** 20 ml/L CPF 6. h, **J:** 20 ml/L CPF 12. h, **K:** 20 ml/L CPF 18. h, **L:** 40 ml/L CPF 6. h, **M:** 40 ml/L CPF 12 h. **N:** 40 ml/L CPF 18. h.

Investigation of Genotoxic Effects of Organophosphorus Pesticides ... 195

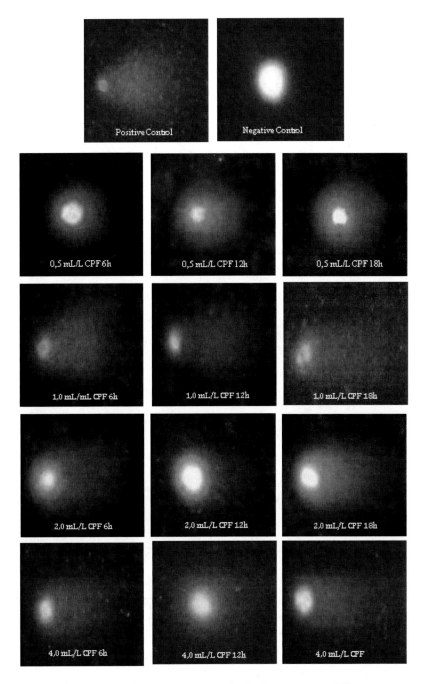

Figure 3. Detection of DNA damage exposure by CPF exposure at different concentrations and times under fluorescence microscope in barley.

Various methods are currently being used to determine whether the DNA molecule is at risk for mutagenic effects and genotoxicity. One of the methods used to examine the genotoxic risk ratios is the Comet test, where direct DNA damage is measured. Single cell gel electrophoresis (SCGE) or so-called "Comet assay"; is a more sensitive and reliable method that results in shorter time than conventional cytogenetic methods used in measuring genotoxic risk ratios (Gichner et al., 2003; Lin et al., 2007). The Comet method has also the advantages such as allowing direct measurement at the DNA level and detecting genetic damage in a short time and directly. Use of this technique also extends the utility of plants in basic and applied studies in environmental mutagenesis. The DNA breaks which migrate out of the nuclei embedded in a thin agarose layer can be measured by comet assay (Menke et al., 2001).

Single cell gel electrophoresis (comet assay) is a quite sensitive and reliable assay resulting in shorter time than common cytogenetic methods used in to analyze the rate of possible genotoxic damage within treated cells. The results of %DNA tail, olive tail moment and tail intensity are the most informative data to detect the DNA damage (Lin et al., 2007; Aydin et al., 2018).

The tail of the comet origins from the migrating relaxed DNA loops from the core in the course of electrophoresis. The distance of DNA loops evaluates the length of the comet tail and it is used to determine the rate of DNA damage. The product of the tail length and the fraction of total DNA in the comet tail are characterized as tail moment. Tail moment includes the smallest detectable comet tail length and intensity of DNA tail. Besides, the tail intensity presents the quantity of DNA breaks and it may be considered as a better predictor of DNA damage than tail length (Collins, 2004; Møller, 2006a,b; Kovačević et al., 2007).

Positive, negative controls of barley seedlings and in the treatment groups (5 ml/L, 10 ml/L, 20 ml/L and 40 ml/L) at the 6., 12. and 18. hours migrated out of the nucleus due to genotoxic damage. Comet formations reflecting the DNA profile were demonstrated by fluorescence staining. When the comet images obtained were examined, significant tail structure was observed in the positive control group (Figure 3b) when compared to the negative control (Figure 3a). Visual comet scores of all results in which the dose and time-dependent genotoxic damage can be observed visually in different CPF administration groups are given in Figure 3C-N.

Pairwise comparisons between the control groups and the negative control group were performed using the Mann-Whitney U test and it was determined that DNA damage was significantly promoted (p <0.001) by CPF application in all groups (Figures 4-5).

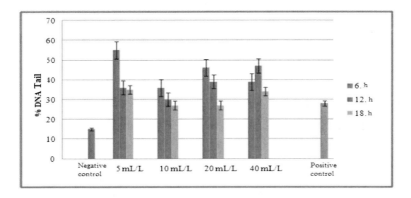

Figure 4. Percentage of DNA damage resulting from CPF exposure in barley plants at different concentrations and times (% DNA Tail).

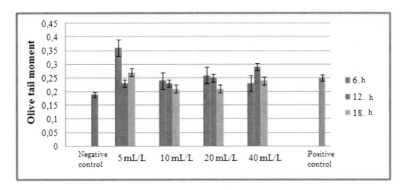

Figure 5. Percentage of DNA damage resulting from CPF exposure in barley plants at different concentrations and times (% DNA Tail moment).

Georgieva and Stoilov (2008) applied different doses of bleomycin to barley root tips, examined single and double strand fractures by comet technique and detected a dose-dependent increase in DNA damage at all doses. 15 minutes after the application they found a reduction in the damage, at the end of the 1st hour after the damage was observed that 50% of the repair. They observed that a more pronounced repair mechanism was effective at the highest dose of bleomycin used. In this study, we observed the highest DNA damage at the end of the 18th hour at 5 ml/L CPF (% DNA tail: 35, Olive tail moment: 0.27). This data supports the information that the repair mechanism is more prominent in increasing doses.

Armalyte and Zukas (2002) determined that 3000 J/m2 and 6000 J/m2 doses of UV damage and repair mechanism by examining the comet technique and they show that increased UV dose due to increased nucleus fragments detected increased in barley root meristem nuclei. After UV exposure the suspensions were kept at room temperature and in the dark after the 2nd and 8th hours, comet test was applied again and DNA damage was decreased in each UV dose. They found a higher repair rate at the end of the 8th hour and suggested that the nucleotide and base excision mechanism was involved in the repair mechanism. In the results of our study, the lowest % DNA tail data of DNA damage after CPF application were obtained at the 18th hour of all doses.

Jovtchev et al. (2001) treated the barley plant with MNU, an alkylating agent, examined DNA damage in leaf and root tip nuclei by comet test and observed a dose-dependent increase. They found that there was a significant decrease in the amount of DNA migration after the 6th hour in the root end nucleus and in the 18th hour in the leaf nucleus. In this study, a significant decrease in DNA damage was detected after 6 hours in leaf nuclei after CPF application on barley seedlings.

Long-term genotoxicity studies can be an important approach to gain insight into the ability of organisms to repair DNA and other protective mechanisms.

Acknowledgement

This research is performed by the support of Research Foundation of Marmara University (BAPKO no. FEN-C-YLP- 031210-0286).

REFERENCES

Armalyte, J., Zukas, K. (2002). Evaluation of UVC-induced DNA damage by SCGE assay and its repair in barley, *Biologija*, 3: 57-59.

Aydin, Y., Uncuoğlu, A.A., Vardar, F. (2018). Single cell gel electrophoresis for determination of plant responses to genotoxic stress. In: Yuksel, B., Karagül, M.S. *Advances in Health and Natural Sciences* Nova Science Publishers, USA, Vol 12.

Collins, A.R., Dobson, V.L., Dusinska, MKennedy, G., Stetina, R. (1997). The comet assay: what can it really tell us?", *Mutation Research*, 375: 183–193.

Collins, A.R. (2004). The comet assay for DNA damage and repair. *Molecular Biotechnology*, 26, 249-261.

Daş, Y.K. (2004). "Investigation some organophosphates and synthetic pyrethroid insecticide residues in honey, produced in Turkey. *PhD Thesis*. Ankara University, Health Sciences Institute.

Dhawan, A., Bajpayee, M., Parmar, D. (2009) Comet assay: a reliable tool for the assessment of DNA damage in different models" *Cell Biol Toxicol.* 25:5-32.

Georgieva, M., Stoilov, L. (2008). Assessment of DNA strand breaks induced by bleomycin in barley by the comet assay, *Environmental and Molecular Mutagenesis*, 49: 381-387.

Gichner, T., Mukherjee, A. (2006).Veleminsky, J.: DNA staining with the fluorochromes EtBr, DAPI and YOYO-1 in the comet assay with tobacco plants after treatment with ethyl methanesulphonate, hyperthermia and Dnase-I. *Mutation Research*, 605: 17-21.

Gichner, T., Patkova, Z., Szakova, J., Znidar, I., Mukherjee, A. (2008b). DNA damage in potato plants induced by cadmium, ethyl methanesulfonate and γ-rays. *Environmental and Experimental Botany*, 62: 113-119.

Gichner, T., Plewa, M.J. (1998).Induction of somatic DNA damage as measured by single cell gel electrophoresis and point mutation in leaves of tobacco plants. *Mutation Research*, 401: 143-152.

Gichner, T., Znidar, I., Szakova, J.(2008a). Evaluation of DNA damage and mutagenicity induced by lead in tobacco plants, *Mutation Research*, 652: 186-190.

Gichner, T., Znider, I., Wagner, E.D., Plewa, M. (2009). The use of higher plants in the comet assay, *The Comet Assay in Toxicology*, Edited by Alok Dhawan and Diana Anderson. *Royal Society of Chemistry*, p: 98- 119.

Gichner, T. (2003). DNA damage induced by indirect and direct acting mutagens in catalase-deficient transgenic tobacco cellular and acellular Comet assays, *Mutation Research* 535, 187-193.

Hayes, W.J., Laws, E.R. (1990). *Handbook of Pesticide Toxicology* Vol. 3, Classes of Pesticides, *Academic Pres*, Inc., NY.

Hoagland, D.R., Arnon, D.I. (1950). The water culture method for growing plants without soil. *Journal Circular,* Berkeley, California, Vol 347.

Jovtchev, G., Menke, M., Schubert, I. (2001). The comet assay detects adaptation to MNU-induced DNA damage in barley, *Mutation Research*, 493: 95-100.

Kovačević, G., Želježic, D., Horvatin, K., Kalafatić, M. (2007). Morphological features and comet assay of green and brown hydra treated with aluminum. *Symbiosis,* 44: 145-152.

Lin, A., Zhang, X., Chen, M., Cao, Q. (2007). Oxidative stress and DNA damages induced by cadmium accumulation, *Journal of Environmental Science*, 19: 596-602.

Lin, A., Zhang, X., Zhu, Y.G., Zhao, F.J. (2008). Arsenate-induced toxicity: effects on antioxidative enzymes and DNA damage in *Vicia faba*, *Environmental Toxicology and Chemistry* 27 (2): 413-419.

Maluszynska, J., Juchimiuk, J. (2005). Plant genotoxicity: a molecular cytogenetic approach in plant bioassays, *Arh. Hig. Rada. Toksikol.,* 56, 177-184.

Mancini, A., Buschini, A., Restivo, F.M., Rossi, C., Poli, P. (2006). Oxidative stress as DNA damage in different transgenic tobacco plants, *Plant Science* 170: 845-852.

Menke, M., Chen, I.P., Angelis K.J., Schubert, I. (2001). DNA damage and repair in Arabidopsis thaliana as measured by the Comet assay after treatment with different classes of genotoxins, *Mutat. Res.,* 493, 87.

Menke, M., Meister, A., Schubert, I. (2000). N-Methyl-N-nitrosourea-induced DNA damage detected by the comet assay in *Vicia faba* nuclei during all interphase stages is not restricted to chromatid aberration hot spots, Mutagenesis 15(6): 503-506.

Moller, P. (2006b). Assessment of reference values for DNA damage detected by the comet assay in human blood cell DNA, *Mutat Res.,* 612: 84-104.

Moller, P. (2006a). The alkaline comet assay: towards validation in biomonitoring of DNA damaging exposures, *Basic Clin. Pharmacol. Toxicol.,* 98: 336-45.

Pourkheirandish, M., Komatsuda, T. (2007). The Importance of Barley Genetics and Domestication in a Global Perspective, *Annals of Botany,* 100 999-1008.

Schardein, J.L., Scialli, A.R. (1999). The legislation of toxicologic safety factors: the Food Quality Protection Act with Chlorpyrifos As a Test Case, *Reproductive Toxicology,* 13(1): 1-14.

Schulte, D., Close, T.J., Graner, A., Langridge, P., Matsumoto, T., Muehlbauer, G., Sato, K., Schulman, A.H., Waugh, R., Wise, R.P., Stein, N. (2009). The International Barley Sequencing Consortium–At the Treshold of Efficient Access to the Barley Genome, *Plant Physiology,* 149: 142- 147.

Steenland, K., Dick, R.B., Howell, R. (2000). Neurologic function among termiticide applicators exposed to chlorpyrifos, *Environmental Health Perspectives,* 108(4): 293-300.

Temel, A., Kartal, G., Gözükırmızı, N. (2008). Genetic and Epigenetic Variations in Barley Calli Cultures, *Biotechnology and Biotechnology Equipment,* 22 (4):911-914.

Thomson, W.T. (1982). Insecticides, acaricides and ovicides, *Agricultural Chemicals,* Book I. Fresno, CA: Thomson Publications.

Tripathi, AK; Pareek, A., Singla-Pareek, S.L. (2016). TUNEL assay to assess extent of DNA fragmentation and programmed cell death in root cells under various stress conditions. *Plant Physioloy*, 7.

Vardar, F., Akgül, N., Aytürk, Ö., Aydın, Y. (2015). Assessment of aluminum induced genotoxicity with comet assay in wheat, rye and triticale roots. *Fresenius Environmental Bulletin*, 37: 3352-3358.

Vardar, F., Çabuk, E., Aytürk, Ö., Aydın, Y. (2016). Determination of aluminum induced programmed cell death characterized by DNA fragmentation in Gramineae species. *Caryologia*, 69: 111-115.

Zhang, X.H., Lin, A.J., Chen, B.D. (2006). Wang, Y.S., Smith, S.E., Smith, F.A.: Effects of Glomus mosseae on the toxicity of heavy metals to *Vicia faba*, *J. Environ. Sci.* (China), 18-4.

In: A Closer Look at the Comet Assay
Editor: Keith H. Harmon

ISBN: 978-1-53611-028-9
© 2019 Nova Science Publishers, Inc.

Chapter 10

GENOTOXICITY OF HYDROQUINONE AND FUNGAL DETOXIFICATION: CORRELATION WITH HYDROQUINONE CONCENTRATION AND CELL VIABILITY

Ana Lúcia Leitão[*]
*Department of Biomass Science and Technology,
Faculty of Science and Technology, New University of Lisbon,
Caparica, Portugal*

ABSTRACT

Hydroquinone, the major benzene metabolite, is an occupational and environmental pollutant. Hydroquinone can be produced as a result of human activities and industrial processes, as well as during phenolic and benzene biotransformation. Additionally, it can be auto-oxidized to form a product, 1,4-benzoquinone, that has higher toxicity than the parent compound. Among fungal strains, the halotolerant *Penicillium chrysogenum* var. *halophenolicum* is a versatile microorganism for hydroquinone biotransformation. The Presto blue® dye assay is a biologically safe and sensitive test for cytotoxic assessment. In the

---

[*] Corresponding Author's Email: aldl@fct.unl.pt.

present study, the correlation between the Comet assay parameters, cell viability, and hydroquinone concentration was performed. It was also investigated the relationship between Comet and remaining hydroquinone after fungal treatment in order to evaluate its biodegradation efficiency. There was a high correlation between %DNA in tail, olive tail moment, tail length and percentage of survival of human colon cancer cells, HCT116, and hydroquinone concentrations. While a positive correlation between Comet parameters and hydroquinone concentrations was achieved; there was a negative correlation between Comet parameters and percentage of survival. *Penicillium chrysogenum* var. *halophenolicum* reduced DNA damage due to hydroquinone degradation. The same trend was observed for experiments under hyperosmolality, reinforcing *Penicillium chrysogenum* var. *halophenolicum* potential for hydroquinone remediation.

Keywords: hydroquinone, genotoxicity, cellular viability, fungal detoxification

INTRODUCTION

Hydroquinone, also known as 1,4-benzenediol, p-benzenediol or p-dihydroxybenzene, the most widely distributed representative of simple phenols, is an aromatic compound consisting of the benzene ring and two – OH groups at the *para* position. In the last decades, hydroquinone like other aromatic compounds sources increased substantially leading to serious toxic effects on fauna, flora, and human. In fact, dietary, smoke, occupational and environment are well established as hydroquinone human exposure sources. Naturally, this phenolic compound could be found as a conjugate with β-D-glucopyranoside in the fruit, bark, and leaves of several plants, especially the ericaceous shrubs such as bearberry, cranberry, as well as in coffee, wheat cereals and broccoli [1]. While in the food industry like red wine, in photography, paints, varnishes, oils motor fuels and cosmetic formulations, hydroquinone has been used.

Hydroquinone is the intermediate of diverse metabolic pathways originated from biological or chemical processes of several aromatics. For example, the biotransformation of benzene in the liver by cytochrome P-

450 monooxygenases converts phenol to hydroquinone, leading its accumulation in the bone marrow [2]. Hydroquinone was also identified as the main degradation product of paracetamol [3]. Meanwhile, in chemical processes such as advanced oxidation processes of aromatics, hydroquinone is one of the compounds produced, as intermediate metabolite of phenol, 4-chlorophenoxyacetic acid and sulfamethoxazole transformation [4-6].

There are an increasing number of evidences that point out hydroquinone as a molecule that triggers apoptosis in multiple cell types [7-10].

In 1993 the cytotoxicity of hydroquinone on mammalian cells was noticed [11], playing Cu(II) an essential role as it promotes the oxidation of hydroquinone with the generation of benzoquinone and reactive oxygen species (ROS) [12]. Meanwhile, genotoxicity and oxidative DNA damage in human hepatoma HepG2 cells independently of the presence of transition metals were reported [13, 14]. Gao et al. support the idea that hydroquinone cytotoxicity exerted over hepatic L02 cells is mediated by intracellular oxidative stress which is responsible for activation of DNA damage [7]. Huang et al. reported that hydroquinone affects the viability of bone marrow-derived mesenchymal stromal cells. The expression of *MDR1* gene, multidrug resistance membrane transporter, is downregulated at the mRNA and protein levels through inhibiting the activation and nuclear translocation of the transcription factor NF-kB [8]. Recently, it was described that hydroquinone induced hematotoxicity in bone marrow mononuclear cells probably due to dysregulation of the Akt/GSK-3β/β-catenin signaling pathway [9].

It is well documented that reactive species originated by hydroquinone could be implicated not only in the formation of modified bases (e.g., 8-oxo-deoxyguanine) in the DNA molecule, which seems to be eliminated with fast kinetics [14], but also in the single and double-stranded DNA breaks [13-15]. Furthermore, hydroquinone stabilizes topoisomerase-mediated DNA scissions since as topoisomerase II poison inhibits the final ligation step of the catalytic cycle of the enzyme [16]. DNA polymerase eta (DNApol η), the product of the xeroderma pigmentosum variant gene, was described to play a relevant role in the response of L-02 cells to hydroquinone induced DNA damage [17]. Taking into account the previous information, it seems that the hydroquinone effect at the DNA level is a serious challenge to genome integrity [18, 19]. According to the last updated recommended lists of genotoxic and non-genotoxic, hydroquinone was included in the group of "*in vivo* genotoxins and/or carcinogens negative or equivocal in Ames and probable aneugens" [20].

Despite the methodologies associated with technological advances, bioremediation is still a green cost-effective and promising strategy for detoxifying phenolic contaminated environments [21, 22]. Among the microorganisms able to grow in the presence of hydroquinone, bacteria like *Burkholderia* sp., *Desulfococcus multivorans*, *Moraxella* sp., *Pseudomonas* sp. [23-26], and fungi like *Cryptococcus* sp., *Trichosporon* sp., *Exophiala jeanselmei*, *Candida* sp., *Geotrichum klebahnii*, *Stephanoascus* sp., *Myxozyma geophila*, *Aspergillus fumigates*, *Candida parapsilosis*, *Tyromyces palustris*, *Gloeophyllum trabeum* and *Phanerochaete chrysosporium* were described [27-32]. On the other hand, the number of reports on biodegradation of hydroquinone by pure culture is scarce, two examples of them were reported using bacteria under anaerobic conditions [23, 33]. Our previous studies have shown that *P. chrysogenum* var. *halophenolicum* (known previously as *Penicillium chrysogenum* CLONA2) uses hydroquinone without previous acclimation under different saline conditions and minimal nutritional requirements [34, 35]. In this work, it is carried out a study to investigate if there is a correlation between genotoxicity and cytotoxicity of hydroquinone in human colon cancer cells (HCT116). It was also analyzed the hydroquinone detoxification potential of *P. chrysogenum* var. *halophenolicum*.

METHODS

Design of Fungal Experiments

P. chrysogenum var. *halophenolicum* (known previously as *Penicillium chrysogenum* CLONA2) was the fungal strain used throughout this study. This fungal strain was previously isolated from a salt mine in Algarve, at the south of Portugal, showing the ability to mineralize phenol under 2% and 5.9% of NaCl [30]. Later, it was characterized based on metabolic features and genetic characteristics as belonging to a subgroup of *P. chrysogenum*, named var. *halophenolicum* [36]. The fungal strain produced non-aromatic natural penicillins rather than benzylpenicillin in a medium containing potassium phenylacetate (the precursor of benzylpenicillin) and was able to grow well on phenylacetic acid using it as a sole carbon source [30, 36].

P. chrysogenum var. *halophenolicum* was maintained at 4 °C on nutrient agar plates supplemented with different saline concentrations (0% and 5.9% (w/v) NaCl) depending on the purpose of the experiment (evaluation of fungal removal capacity at 0% or 5.9% (w/v) NaCl). Batch experiments were performed in an orbital shaker (Certomat® BS-T Incubator, Sartorius Stedim Biotech, Goettingen, Germany) at 160 rpm and 25±1 °C under dark conditions to prevent hydroquinone photo-oxidation. Three replicates were used per test assayed. Uninoculated control flasks (duplicates) were incubated and aerated in parallel as negative controls of the experiment.

Precultures of cells were routinely aerobically cultivated in MC medium (30.0 g L^{-1} glucose, 3.0 g L^{-1} $NaNO_3$, 0.5 g L^{-1} $MgSO_4.7H_2O$, 10 mg L^{-1} $NH_4Fe(SO_4)_2.12H_2O$, 1.0 g L^{-1} K_2HPO_4, 5.0 g L^{-1} Yeast extract, and supplemented or not with 58.5 g L^{-1} NaCl depending on the experiment; pH was adjusted to 5.6-5.8 with 5 mol L^{-1} HCl). Then, cells were centrifuged for 10 min at 10,000 x *g* and washed three times in 0.85% (w/v) of NaCl. The pellet (10%) was inoculated in MMFe medium (1 g L^{-1} K_2HPO_4, 1 g L^{-1} $(NH_4)_2SO_4$, 200 mg L^{-1} $MgSO_4.7H_2O$, 33 mg L^{-1} $FeCl_3.6H_2O$, 100 mg L^{-1} $CaCl_2$, and supplemented or not with 58.5 g L^{-1}

NaCl depending on the experiment; pH was adjusted to 5.6-5.8 with 5 mol L^{-1} HCl) with 600 mg L^{-1} of hydroquinone.

Hydroquinone concentration was monitored up to an incubation time which allowed its concentration to values below to 2 mg/L. Hydroquinone concentration was quantified by High-Performance Liquid Chromatography apparatus L-7100 (LaChrom HPLC System, Merck) consisting of a quaternary pump system, and L-7400 UV detector. The whole system was controlled using HPLC System Manager software for Windows NT™. The separation of the analytes was performed with a LiChrocart 250-4 RT-18 end capped (5 µm) column (Merck, Germany), using an isocratic condition (mobile phase: water: acetonitrile (70:30 (v/v)), and a flow rate of 1.0 mL/min. Detection was performed at 254 nm. Hydroquinone could be separated and concentrations estimated within 10 min, using standard.

Cell Culture and Viability Assays

Adherent colon cancer HCT116 cells (ATCC CCL-247) were cultured on 75 cm^2 T-flasks in McCoy's 5a modified medium supplemented with 10% heat-inactivated fetal bovine serum, 2 mM L-glutamine, 1% MEM non-essential amino acids and 100 U/mL penicillin/streptomycin (Gibco, Life Technologies), and maintained at 37°C in a humidified incubator under 6% CO$_2$. Cells were plated at a density of 1 x 10^4 cells/well in 24-well plates and incubated for 24 h before initiation of experiments using McCoy's supplemented with different concentrations of hydroquinone (1 mg L^{-1}, 5 mg L^{-1}, 10 mg L^{-1}, 25 mg L^{-1} and 50 mg L^{-1}) in the case of assays for evaluate hydroquinone effect on HCT116 cells; while, in the experiments for assessing bioremediation potential of *P. chrysogenum* var. *halophenolicum* McCoy's was supplemented either (1) MMFe medium originating from cultures of *P. chrysogenum* var. *halophenolicum* after hydroquinone depletion, i.e., at a concentration below to 2 mg/L (conditioned composite medium), (2) freshly prepared MMFe medium (plain composite medium), or (3) medium supplemented with 600 mg L^{-1} hydroquinone.

Cell viability was assessed using Presto Blue® reagent (Thermo Fisher Scientific), a commercial assay which is based on the reduction of the cell permeable redox indicator resazurin (deep blue) into resorufin (red and fluorescent) by viable, metabolically active cells. At the end of specified incubation times, 40 µL of Presto Blue® 10X solution was added per 1 mL of culture medium and incubated for an additional time of 1 h. Fluorescence emission was detected in a Tecan Infinite M200 plate reader, using an excitation wavelength of 535 nm and an emission wavelength of 615 nm, as recommended by the manufacturer. Results were processed and analyzed by the Tecan i-Control v. 1.4.5.0 plate reader software. Each experiment was performed in technical triplicates.

Alkaline Single Cell Gel Electrophoresis: Comet Assay

DNA damage represented by strand breaks was evaluated by alkaline single-cell electrophoresis using Trevigen Comet Assay® kit (Trevigen Inc., Gaithersburg, MD, USA). Selected cell samples were trypsinized and further resuspended in ice-cold PBS (Ca^{2+} and Mg^{2+} free) to a concentration of 1 × 10^5 cells/mL. An aliquot of 5 µL of cells was added to 50 µL of molten LM Agarose (1% low-melting agarose) kept at 37 °C. For the assay, 50 µL of the agarose-cell suspension were pipetted immediately and carefully spread onto the comet slides. To accelerate the gelling of the agarose discs, slides were incubated at 4°C in the dark for 10 min and then transferred to prechilled lysis solution (2.5 M NaCl, 100 mM EDTA, 10 mM Tris-base, containing 1% sodium lauryl sarcosinate and 1% Triton X-100, pH 10) for 30 min at 4°C. A subsequent denaturation step was performed in alkaline conditions (300 mM NaOH, 1 mM EDTA, pH > 13) at room temperature for 30 min, incubating the slides in the dark. After denaturation, slides were transferred to prechilled alkaline electrophoresis solution pH > 13 (300 mM NaOH, 1 mM EDTA) and subjected to electrophoresis at 1 V/cm, 300 mA for 30 min in the dark at 4 °C. The slides were then rinsed with deionized water three times and immersed in 70% ethanol at room temperature for 5 min and let dry for 10 additional minutes. DNA was stained with 100 µL of SYBR Green I dye (Trevigen, 1:10,000 in Tris–EDTA buffer, pH 7.5) for 10 min at 4 °C in the dark and immediately analyzed using a CCD camera (Roper Scientific Coolsnap HQ CCD) attached to a Zeiss Axiovert 200M widefield fluorescence microscope. Comets were visualized with an excitation filter of 450–490 nm and an emission filter of 515 nm and fluorescent images of single cells were captured at 200X magnification. For statistical purposes, a minimum of 100 randomly chosen cells per experimental group were scored for comet parameters including tail length, percentage of DNA in tail and olive tail moment using the Tritek CometScore Freeware v1.5 image analysis software, and following the already described protocol [37].

Statistical Analysis

The correlation between individual variables was calculated using Spearman correlation coefficient; p values <0.05 were considered statistically significant.

RESULTS AND DISCUSSION

Terrestrial and aquatic species may be exposed to a complex array of pollutants, including aromatic compounds and their intermediates metabolites such as hydroquinone [19, 20]. Taking into account environmentally friendly and cost-effective technology, there is an urgent need for the search of organisms that could be protagonist in remediation of contaminated sites, although the difficulties in extrapolating laboratory derived data to the field.

The cytotoxic effect of hydroquinone has been previously assessed using a range of concentrations from 25 mg L^{-1} to 500 mg L^{-1} [35]. It was shown that hydroquinone treatment reduced the viability of colon cancer

HCT116 cells in a dose-dependent manner. This effect was particularly achieved at high concentrations of hydroquinone (250 mg L^{-1} and 500 mg L^{-1}) [35]. Since the present work aims to assess the potential of *P. chrysogenum* var. *halophenolicum* to remediate contaminated sites, the hydroquinone effect in HCT116 cells at lower hydroquinone levels was investigated.

Table 1 shows correlations between comet parameters and hydroquinone concentrations as well as cell viability. Hydroquinone and genotoxicity were highly correlated. There were significant relationships between hydroquinone and %DNA in tail, tail length and olive tail moment (p = 0.012, 0.007 and 0.010, respectively), indicating a positive association between DNA strand breakages and hydroquinone levels. While, a significant negative correlation between levels of hydroquinone and percentage of survival, Spearman correlation of -0.935 (p = 0.006), was obtained. Finally, cell viability and genotoxicity were strongly negatively correlated. It was found a significant negative correlation between cell viability and % DNA in tail, tail length and olive tail moment (p = $2.15e^{-004}$, $3.12e^{-006}$ and $8.20e^{-005}$, respectively). This finding was expected since high DNA damage can lead to a reduction of cell viability [38].

Table 1. The correlations among the genotoxicity, hydroquinone and cellular viability

	Hydroquinone	% Survival
% DNA in tail	0.909 (p = 0.012)	-0.988 (p = $2.15e^{-004}$)
Tail length	0.929 (p = 0.007)	-0.999 (p = $3.12e^{-006}$)
Olive tail moment	0.919 (p = 0.010)	-0.993 (p = $8.20e^{-005}$)

Recently, it was noticed that phenolic compounds have cytotoxic and genotoxic effect on cancer cells [39, 40]. For example, carvacrol, thymol and their mixture showed to induce cytotoxicity and genotoxicity on the human gastric carcinoma, decreasing GSH levels and increasing ROS [39]. In fact, in human leukemia line cell, HL-60 cells, reactive oxygen species was the probable mechanism by which hydroquinone induces HL-60 cells to enter the cell cycle and proliferate [41].

To assess hydroquinone removal capacity by *P. chrysogenum* var. *halophenolicum*, batch cultures were performed in liquid media (MMFe medium) with (5.9% NaCl) and without salt supplemented with 600 mg L^{-1} of hydroquinone as the sole carbon and energy source. The fungal strain was able to remove hydroquinone to concentrations below 2 mg L^{-1} in both systems. However, the removal rate was higher under no salt than in corresponding NaCl treatments. In fact, in absence of NaCl *P. chrysogenum* var. *halophenolicum* was able to remove more than 98% of hydroquinone in 68 hours, while more than 96 hours were needed to achieve the same goal at 5.9% NaCl.

To further investigate the fungal potential in remediation field, DNA damage experiments were performed using the comet assay. DNA damage is defined as a modification of DNA structure by internal or external cellular stress able of causing cellular injury and reduction of viability or reproductive success of the organism [38].

Comet dataset was visualized in a 3D scatter plot graph, as shown in Figure 1. The graphs showed a clear separation of the different conditions independently of the presence of salt, indicating that the genotoxic effect of the hydroquinone in HCT116 cells differs from treated to untreated fungal samples. In both conditions it was shown that the presence of *P. chrysogenum* var. *halophenolicum* not only had the ability to reduce hydroquinone levels but also decreased the toxicity of the system.

The olive tail moment for HCT116 cells was increased in HQ exposed samples compared to the control. However, when fungal strain used hydroquinone, the olive tail moment was smaller compared with the untreated samples (Figure 2). In fact, cellular viability was higher than 90% in the samples treated with the *P. chrysogenum* var. *halophenolicum*, which are probably due to the production of no-toxic and/or less toxic metabolites than the original compounds.

The present results indicating that the fungal strain decreased not only hydroquinone levels but also the toxicity. Although, fungal treated samples from cultures under 5.9% NaCl showed to be less effective. According to several studies, hyperosmolality in form of high NaCl caused DNA damage [42,43]. For example in rabbit renal inner medullary epithelial cells [44], Chinese hamster ovary cells [45] and murine kidney cells [43], DNA damage by high osmotic stress was observed due to inhibition of DNA repair mechanisms or an increase in the frequency of DNA double-strand breaks [43].

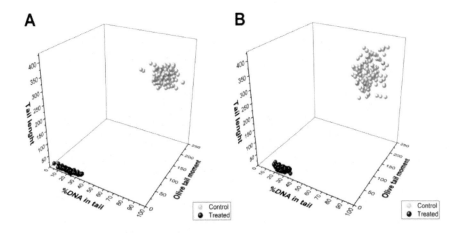

Figure 1. Tridimensional scatter plot of comet parameters (Tail length, % of DNA in tail and olive tail moments) obtained from HCT116 cells incubated with culture supernatants from *P. chrysogenum* var. *halophenolicum* containing hydroquinone before (Control) and after (Treated) treatment with the fungal strain. Each depicted symbol represents a single cell experiment. **A**, comet parameters obtained from samples where the fungal strain was cultivated on media without salt; **B**, comet parameters obtained from samples where the fungal strain was cultivated on media supplemented with 5.9% NaCl.

In summary, hydroquinone induced DNA damage to HCT116 cells in the form of DNA double-strand breaks as it was demonstrated by alkaline comet assay. Genotoxicity was highly correlated with hydroquinone and negatively correlated with cell viability. *P. chrysogenum* var. *halophenolicum* had the capacity, without prior acclimation, to remove hydroquinone present at high initial concentrations to levels that are non-genotoxic to human cells. Furthermore, the fungal efficacy of reduction DNA damage was lower under saline conditions, probably due to NaCl toxicity, showing its potential for remediation of hydroquinone in hypersaline contaminated environments.

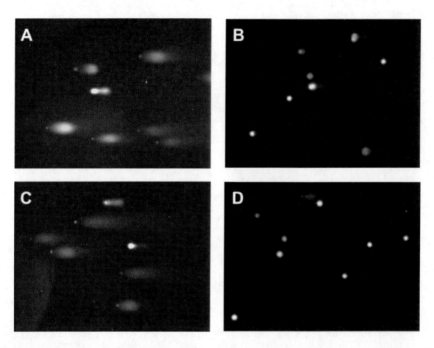

Figure 2. Selected microscopic fields from alkaline single-cell electrophoresis experiments from HCT116 cells incubated with culture supernatants from *P. chrysogenum* var. *halophenolicum* on media containing hydroquinone. Panels A and B depict the shape of nuclei of HTC116 after electrophoresis incubated with fungal culture supernatants obtained in culture media without salt, before and after fungal treatment, respectively. Panels C and D depict the shape of nuclei of HTC116 after electrophoresis incubated with fungal culture supernatants obtained in culture media supplemented with 5.9% NaCl, before and after fungal treatment, respectively.

ACKNOWLEDGMENTS

Author thanks to Prof. Francisco Enguita, Department of Cellular Biology, University of Medicine for his valuable scientific contribution and to MEtRICs, ref. UID/EMS/04077/2019.

REFERENCES

[1] Deisinger, P. J.; Hill, T. S.; English, J. C. Human exposure to naturally occurring hydroquinone. *J. Toxicol. Environ. Health* 1996, *47*, 31-46, doi:10.1080/009841096161915.

[2] Subrahmanyam, V. V.; Kolachana, P.; Smith, M. T. Hydroxylation of phenol to hydroquinone catalyzed by a human myeloperoxidase-superoxide complex: possible implications in benzene-induced myelotoxicity. *Free Radic. Res. Commun.* 1991, *15*, 285-296.

[3] Zur, J.; Wojcieszynska, D.; Hupert-Kocurek, K.; Marchlewicz, A.; Guzik, U. Paracetamol-toxicity and microbial utilization. *Pseudomonas moorei* KB4 as a case study for exploring degradation pathway. *Chemosphere* 2018, *206*, 192-202, doi: 10.1016/j.chemosphere.2018.04.179.

[4] Boye, B.; Dieng, M.M.; Brillas, E. Degradation of herbicide 4-chlorophenoxyacetic acid by advanced electrochemical oxidation methods. *Environ. Sci. Technol.* 2002, *36*, 3030-3035.

[5] Kusic, H.; Koprivanac, N.; Bozic, A. L.; Selanec, I. Photo-assisted Fenton type processes for the degradation of phenol: a kinetic study. *J. Hazard. Mater.* 2006, *136*, 632-644, doi:10.1016/j.jhazmat.2005.12.046.

[6] Mulla, S. I.; Hu, A.; Sun, Q.; Li, J.; Suanon, F.; Ashfaq, M.; Yu, C. P. Biodegradation of sulfamethoxazole in bacteria from three different origins. *J. Environ. Manage.* 2018, *206*, 93-102, doi:10.1016/j.jenvman.2017.10.029.

[7] Gao, Y.; Tang, H.; Xiong, L.; Zou, L.; Dai, W.; Liu, H.; Hu, G. Protective Effects of Aqueous Extracts of Flos *Lonicerae japonicae* against Hydroquinone-Induced Toxicity in Hepatic L02 Cells. *Oxid. Med. Cell. Longev.* 2018, *2018*, 4528581, doi:10.1155/2018/4528581.

[8] Huang, J.; Zhao, M.; Li, X.; Ma, L.; Zhang, J.; Shi, J.; Li, B.; Fan, W.; Zhou, Y. The Cytotoxic Effect of the Benzene Metabolite Hydroquinone is Mediated by the Modulation of MDR1 Expression

via the NF-kappaB Signaling Pathway. *Cell. Physiol. Biochem.* 2015, *37*, 592-602, doi:10.1159/000430379.
[9] Li, J.; Jiang, S.; Chen, Y.; Ma, R.; Chen, J.; Qian, S.; Shi, Y.; Han, Y.; Zhang, S.; Yu, K. Benzene metabolite hydroquinone induces apoptosis of bone marrow mononuclear cells through inhibition of beta-catenin signaling. *Toxicol. In Vitro* 2018, *46*, 361-369, doi:10.1016/j.tiv.2017.08.018.
[10] Xu, L.; Liu, J.; Chen, Y.; Yun, L.; Chen, S.; Zhou, K.; Lai, B.; Song, L.; Yang, H.; Liang, H., et al. Inhibition of autophagy enhances Hydroquinone-induced TK6 cell death. *Toxicol. In Vitro* 2017, *41*, 123-132, doi:10.1016/j.tiv.2017.02.024.
[11] Li, Y.; Trush, M. A. DNA damage resulting from the oxidation of hydroquinone by copper: role for a Cu(II)/Cu(I) redox cycle and reactive oxygen generation. *Carcinogenesis* 1993, *14*, 1303-1311.
[12] Li, Y.; Trush, M. A. Oxidation of hydroquinone by copper: chemical mechanism and biological effects. *Arch. Biochem. Biophys.* 1993, *300*, 346-355, doi:S0003986183710477.
[13] Luo, L.; Jiang, L.; Geng, C.; Cao, J.; Zhong, L. Hydroquinone-induced genotoxicity and oxidative DNA damage in HepG2 cells. *Chem. Biol. Interact.* 2008, *173*, 1-8, doi:10.1016/j.cbi.2008.02.002.
[14] Peng, C.; Arthur, D.; Liu, F.; Lee, J.; Xia, Q.; Lavin, M.F.; Ng, J.C. Genotoxicity of hydroquinone in A549 cells. *Cell Biol. Toxicol.* 2013, *29*, 213-227, doi:10.1007/s10565-013-9247-0.
[15] Ishihama, M.; Toyooka, T.; Ibuki, Y. Generation of phosphorylated histone H2AX by benzene metabolites. *Toxicol. In Vitro* 2008, *22*, 1861-1868, doi:10.1016/j.tiv.2008.09.005.
[16] Lindsey, R. H., Jr.; Bender, R. P.; Osheroff, N. Effects of benzene metabolites on DNA cleavage mediated by human topoisomerase II alpha: 1,4-hydroquinone is a topoisomerase II poison. *Chem. Res. Toxicol.* 2005, *18*, 761-770, doi:10.1021/tx049659z.
[17] Hu, G.; Huang, H.; Yang, L.; Zhong, C.; Xia, B.; Yang, Y.; Liu, J.; Wu, D.; Liu, Q.; Zhuang, Z. Down-regulation of Poleta expression leads to increased DNA damage, apoptosis and enhanced S phase

arrest in L-02 cells exposed to hydroquinone. *Toxicol. Lett.* 2012, *214*, 209-217, doi:10.1016/j.toxlet.2012.08.025.

[18] Barreto, G.; Madureira, D.; Capani, F.; Aon-Bertolino, L.; Saraceno, E.; Alvarez-Giraldez, L. D. The role of catechols and free radicals in benzene toxicity: an oxidative DNA damage pathway. *Environ. Mol. Mutagen.* 2009, *50*, 771-780, doi:10.1002/em.20500.

[19] Enguita, F. J.; Leitao, A. L. Hydroquinone: environmental pollution, toxicity, and microbial answers. *Biomed. Res. Int.* 2013, *2013*, 542168, doi:10.1155/2013/542168.

[20] Kirkland, D.; Kasper, P.; Martus, H.J.; Muller, L.; van Benthem, J.; Madia, F.; Corvi, R. Updated recommended lists of genotoxic and non-genotoxic chemicals for assessment of the performance of new or improved genotoxicity tests. *Mutat. Res. Genet. Toxicol. Environ. Mutagen.* 2016, *795*, 7-30, doi:10.1016/j.mrgentox.2015.10.006.

[21] Marco-Urrea, E.; Garcia-Romera, I.; Aranda, E. Potential of non-ligninolytic fungi in bioremediation of chlorinated and polycyclic aromatic hydrocarbons. *N. Biotechnol.* 2015, *32*, 620-628, doi:10.1016/j.nbt.2015.01.005.

[22] Quintella, C. M.; Mata, A. M. T.; Lima, L. C. P. Overview of bioremediation with technology assessment and emphasis on fungal bioremediation of oil contaminated soils. *J. Environ. Manage.* 2019, *241*, 156-166, doi:10.1016/j.jenvman.2019.04.019.

[23] Schnell, S.; Bak, F.; Pfennig, N. Anaerobic degradation of aniline and dihydroxybenzenes by newly isolated sulfate-reducing bacteria and description of *Desulfobacterium anilini*. *Arch. Microbiol.* 1989, *152*, 556-563.

[24] Arora, P. K.; Jain, R. K. Metabolism of 2-chloro-4-nitrophenol in a gram negative bacterium, *Burkholderia* sp. RKJ 800. *PLoS One* 2012, *7*, e38676, doi:10.1371/journal.pone.0038676.

[25] Spain, J. C.; Gibson, D. T. Pathway for Biodegradation of p-Nitrophenol in a Moraxella sp. *Appl. Environ. Microbiol.* 1991, *57*, 812-819.

[26] Zhang, J. J.; Liu, H.; Xiao, Y.; Zhang, X. E.; Zhou, N. Y. Identification and characterization of catabolic para-nitrophenol 4-

monooxygenase and para-benzoquinone reductase from *Pseudomonas* sp. strain WBC-3. *J. Bacteriol.* 2009, *191*, 2703-2710, doi:10.1128/JB.01566-08.
[27] Eppink, M. H.; Cammaart, E.; Van Wassenaar, D.; Middelhoven, W. J.; van Berkel, W. J. Purification and properties of hydroquinone hydroxylase, a FAD-dependent monooxygenase involved in the catabolism of 4-hydroxybenzoate in *Candida parapsilosis* CBS604. *Eur. J. Biochem.* 2000, *267*, 6832-6840, doi:ejb1783.
[28] Jones, K. H.; Trudgill, P. W.; Hopper, D. J. 4-Ethylphenol metabolism by *Aspergillus fumigatus*. *Appl. Environ. Microbiol.* 1994, *60*, 1978-1983.
[29] Kamada, F.; Abe, S.; Hiratsuka, N.; Wariishi, H.; Tanaka, H. Mineralization of aromatic compounds by brown-rot basidiomycetes - mechanisms involved in initial attack on the aromatic ring. *Microbiology* 2002, *148*, 1939-1946.
[30] Leitão, A. L.; Duarte, M. P.; Santos Oliveira, J. Degradation of phenol by a halotolerant strain of *Penicillium chrysogenum*. *Int. Biodeterior. Biodegrad.* 2007, *59*, 220-225.
[31] Nakamura, T.; Ichinose, H.; Wariishi, H. Flavin-containing monooxygenases from *Phanerochaete chrysosporium* responsible for fungal metabolism of phenolic compounds. *Biodegradation* 2012, *23*, 343-350, doi:10.1007/s10532-011-9521-x.
[32] Middelhoven, W. J. Catabolism of benzene compounds by ascomycetous and basidiomycetous yeasts and yeastlike fungi. A literature review and an experimental approach. *Antonie Van Leeuwenhoek* 1993, *63*, 125-144.
[33] Szewzyk, U.; Schink, B. Degradation of hydroquinone, gentisate, and benzoate by a fermenting bacterium in pure or defined mixed culture. *Arch. Microbiol.* 1989, *151*, 541-545.
[34] Guedes, S. F.; Leitão, A. L. Simultaneous removal of dihydroxybenzenes and toxicity reduction by *Penicillium chrysogenum* var. *halophenolicum* under saline conditions. *Ecotoxicol. Environ. Saf.* 2018, *150*, 240-250.

[35] Pereira, P.; Enguita, F. J.; Ferreira, J.; Leitão, A. L. DNA damage induced by hydroquinone can be prevented by fungal detoxification. *Toxicol. Rep.* 2014, *1*, 1096-1105.

[36] Leitão, A. L.; Garcia-Estrada, C.; Ullan, R. V.; Guedes, S. F.; Martin-Jimenez, P.; Mendes, B.; Martin, J. F. *Penicillium chrysogenum* var. *halophenolicum*, a new halotolerant strain with potential in the remediation of aromatic compounds in high salt environments. *Microbiol. Res.* 2012, *167*, 79-89, doi:10.1016/j.micres.2011.03.004.

[37] Lovell, P. D.; Omori, T. Statistical issues in the use of the comet assay. *Mutagenesis* 2008, *23*, 171-182.

[38] Kaufmann, W. K.; Paules, R. S. DNA damage and cell cycle checkpoints. *FASEB J.* 1996, *10*, 238-247, doi:10.1096/fasebj.10.2.8641557.

[39] Gunes-Bayir, A.; Kocyigit, A.; Guler, E. M. In vitro effects of two major phenolic compounds from the family Lamiaceae plants on the human gastric carcinoma cells. *Toxicol. Ind. Health* 2018, *34*, 525-539, doi:10.1177/0748233718761698.

[40] Koyuncu, I.; Gonel, A.; Akdag, A.; Yilmaz, M. A. Identification of phenolic compounds, antioxidant activity and anti-cancer effects of the extract obtained from the shoots of *Ornithogalum narbonense* L. *Cell Mol. Biol. (Noisy-le-grand)* 2018, *64*, 75-83, doi:10.14715/cmb/2018.64.1.14.

[41] Ruiz-Ramos, R.; Cebrian, M. E.; Garrido, E. Benzoquinone activates the ERK/MAPK signaling pathway via ROS production in HL-60 cells. *Toxicology* 2005, *209*, 279-287, doi:10.1016/j.tox.2004.12.035.

[42] Dmitrieva, N. I.; Burg, M. B. Living with DNA breaks is an everyday reality for cells adapted to high NaCl. *Cell Cycle* 2004, *3*, 561-563, doi:869 [pii].

[43] Kultz, D.; Chakravarty, D. Hyperosmolality in the form of elevated NaCl but not urea causes DNA damage in murine kidney cells. *Proc. Natl. Acad. Sci. USA.* 2001, *98*, 1999-2004, doi:10.1073/pnas.98.4.1999.

[44] Uchida, S.; Green, N.; Coon, H.; Triche, T.; Mims, S.; Burg, M. High NaCl induces stable changes in phenotype and karyotype of renal cells in culture. *Am. J. Physiol.* 1987, *253*, C230-242, doi:10.1152/ajpcell.1987.253.2.C230.

[45] Galloway, S. M.; Deasy, D. A.; Bean, C. L.; Kraynak, A. R.; Armstrong, M. J.; Bradley, M. O. Effects of high osmotic strength on chromosome aberrations, sister-chromatid exchanges and DNA strand breaks, and the relation to toxicity. *Mutat. Res.* 1987, *189*, 15-25, doi:0165-1218(87)90029-2.

INDEX

A

acid, 4, 5, 9, 10, 11, 24, 114, 115, 119, 145, 153, 163, 167, 168, 205, 207, 215
adaptation, 165, 184, 200
adverse effects, 51, 110, 143, 145, 146, 151
agriculture, 40, 109, 137, 145, 171, 173, 175, 188, 189
aluminum toxicity, 144, 162, 164, 165, 166, 168, 169, 170
animal studies, 4, 24, 126
anthocyanin, 146, 154, 155
antioxidant, 5, 11, 12, 15, 100, 128, 146, 148, 155, 158, 162, 163, 164, 219
apoptosis, 41, 58, 108, 126, 140, 141, 205, 216
aromatic compounds, 204, 210, 218, 219
arrest, 130, 141, 217
assessment, 39, 49, 52, 55, 58, 86, 92, 106, 107, 126, 130, 131, 175, 183, 185, 199, 203, 217

B

bacteria, 206, 215, 217
bacterium, 217, 218

base 4, 6, 10, 12, 19, 86, 87, 92, 98, 112, 114, 127, 198, 209
benzene, 5, 203, 204, 215, 216, 217, 218
biodegradation, 174, 204, 206
biological systems, 37, 48
biomarkers, 22, 59, 129, 134, 162, 186
biomedical applications, 48, 50
biomonitoring, 2, 9, 10, 12, 13, 17, 22, 25, 27, 28, 29, 30, 34, 86, 92, 93, 99, 105, 136, 201
biopsy, 89, 90, 97
bioremediation, 206, 208, 217
biotic, 173, 174, 175, 190
bladder cancer, 8, 9, 18, 32, 96
blood, 3, 5, 8, 13, 14, 17, 21, 23, 25, 26, 34, 89, 108, 113, 115, 116, 120, 126, 135, 166, 201
bone, 8, 110, 134, 139, 205, 216
bone marrow, 8, 110, 139, 205, 216
brain, 4, 40, 42
breakdown, 146, 151, 155, 156
bronchial epithelial cells, 54, 56, 62

C

Ca^{2+}, 146, 147, 209
cadmium, 164, 165, 166, 184, 199, 200

Calypso, 106, 110, 112, 116, 120, 121, 122, 123, 124, 125, 127, 130
cancer, 2, 3, 4, 5, 7, 8, 9, 11, 12, 14, 15, 16, 18, 21, 22, 23, 24, 29, 30, 32, 46, 49, 51, 58, 59, 80, 85, 86, 88, 89, 90, 94, 96, 97, 100, 101, 102, 106, 125, 131, 135, 208, 210, 211, 219
cancer cells, 89, 211
carbon, 48, 50, 54, 59, 62, 64, 87, 207, 212
carbon nanotubes, 48, 54, 62, 64
carcinogenesis, 12, 13, 89, 90
carcinoma, 56, 64, 211, 219
cell culture, 22, 36, 45, 46, 51, 106, 131
cell cycle, 89, 109, 141, 212, 219
cell death, 53, 146, 155, 160, 166, 169, 170, 202, 216
cellular viability, 117, 129, 203, 204, 209, 211, 212, 213
chemical, 6, 11, 14, 18, 24, 37, 48, 57, 107, 109, 130, 139, 173, 174, 175, 188, 204, 216
chlorophyll, 144, 148, 153, 154, 165, 170
Chlorpyrifos, 134, 187, 188, 189, 201
chromatin loops, 66, 67, 73, 74, 75, 76, 78, 79, 81, 82, 108
chromosome, 82, 83, 110, 135, 146, 181, 220
CO_2, 151, 167, 208
cohesin, 74, 76, 82
CometChip, 63, 94, 100
commercial, 46, 106, 110, 111, 113, 120, 121, 124, 130, 134, 139, 186, 209
complications, 9, 11, 38
composition, 27, 37, 119, 134, 146
compounds, 2, 14, 15, 25, 47, 86, 175, 181, 189, 205, 212, 218
contaminated soil, 176, 185, 217
contamination, 115, 125, 139
control group, 25, 27, 149, 159, 178, 179, 193, 196, 197
cooperativity of DNA exit, 69

correlation, 7, 8, 12, 14, 24, 25, 26, 42, 43, 45, 77, 168, 179, 203, 204, 206, 210, 211
CpG islands, 89
CTCF, 74, 76, 82, 83
cultivation, 106, 112, 115, 116, 117, 119, 121, 124, 125, 128, 130
culture, 46, 52, 112, 115, 116, 124, 129, 130, 131, 135, 165, 184, 200, 206, 209, 213, 214, 218, 220
culture media, 112, 129, 214
culture medium, 46, 115, 116, 209
cytotoxic, 11, 22, 43, 44, 50, 51, 57, 62, 110, 111, 131, 132, 136, 139, 141, 203, 210, 211, 215
cytotoxicity, 46, 57, 58, 61, 62, 138, 140, 205, 206, 211

D

damages, 32, 60, 65, 66, 79, 92, 166, 184, 185, 186, 200
degradation, 82, 88, 156, 158, 174, 178, 188, 204, 205, 215, 217, 218
detection, 19, 21, 34, 35, 36, 54, 66, 86, 89, 90, 91, 95, 96, 97, 105, 106, 107, 108, 130, 168, 171, 172, 183
detoxification, 155, 204, 206, 219
developing countries, 18, 24, 31
diabetes, 5, 7, 9, 11, 101, 125, 139
diseases, 2, 7, 25, 86, 106, 108, 173
distribution, 57, 75, 76, 77, 136, 155, 164, 171, 174, 175
DNA amount in the tails, 67, 70, 72, 74
DNA lesions, 29, 41, 92, 108, 140
DNA loops, 36, 65, 66, 70, 74, 75, 79, 82, 108, 158, 196
DNA methylation, 85, 86, 87, 88, 90, 91, 92, 93, 94, 97, 98, 99, 102
DNA repair, 2, 5, 8, 10, 14, 36, 58, 81, 88, 92, 99, 130, 134, 138, 176, 213
DNA strand breaks, 17, 31, 32, 35, 86, 93, 106, 140, 184, 199, 220

drugs, 18, 22, 34, 126, 188

E

ecosystem, 164, 171, 175, 187, 189
electrophoresis, 2, 3, 13, 14, 17, 19, 20, 21, 23, 27, 30, 32, 34, 36, 37, 39, 65, 66, 67, 68, 69, 70, 71, 72, 75, 78, 80, 81, 91, 93, 107, 114, 115, 118, 131, 135, 136, 144, 147, 149, 158, 162, 172, 175, 177, 183, 184, 188, 191, 192, 196, 199, 209, 214
emission, 209, 210
Endosulfan (ES), 171, 172, 173, 174, 176, 177, 178, 179, 180, 181, 182, 183, 184, 185, 188
environment, 2, 18, 20, 27, 37, 48, 96, 125, 174, 186, 189, 204
environmental factors, 87, 145
environmental stress, 155, 163, 169
enzyme, 52, 93, 94, 128, 144, 146, 148, 149, 153, 155, 158, 205
enzymes, 3, 36, 44, 128, 153, 156, 158, 162, 163, 170, 185, 193, 200
EpiComet, 86, 93, 94, 95, 99
EpiComet-Chip, 86, 94, 95, 99
epidemiology, 32, 86, 98, 175
epidermis, 151, 152, 168
epigenetic, 85, 87, 88, 96, 101, 102, 201
epithelial cells, 3, 21, 41, 43, 59, 213
erythrocytes, 111, 115, 116, 133
ethylene, 114, 115, 163

F

false negative, 130
false positive, 97, 129
fluorescence, 3, 19, 36, 38, 67, 73, 92, 95, 107, 149, 153, 165, 166, 180, 193, 195, 196, 210
food chain, 171, 173, 174, 175

formation, 27, 52, 65, 66, 67, 68, 69, 70, 80, 81, 88, 108, 110, 111, 128, 145, 152, 174, 205
fractures, 19, 20, 197
fragments, 20, 66, 80, 89, 108, 144, 187, 198
free radicals, 155, 164, 217
friction, 69, 71, 72
functionalization, 47, 50, 54, 62
fungal detoxification, 203, 204, 219
fungi, 176, 190, 206, 217, 218

G

gel, 2, 13, 14, 17, 19, 23, 30, 31, 32, 33, 34, 36, 38, 59, 66, 67, 80, 81, 91, 118, 135, 136, 146, 149, 158, 162, 172, 175, 177, 184, 191, 192, 196, 199
gene expression, 86, 87, 88, 126
gene regulation, 85, 96
genes, 3, 87, 88, 89, 96, 97, 107, 140
genetics, 12, 13, 14, 15, 87
genome, 76, 83, 85, 88, 91, 99, 126, 130, 206
genome-wide hypomethylation, 89
genotoxicity, 2, 4, 15, 18, 19, 28, 33, 35, 36, 39, 40, 41, 42, 43, 44, 45, 46, 47, 48, 49, 50, 52, 53, 54, 55, 56, 57, 58, 59, 61, 62, 63, 64, 80, 86, 92, 94, 105, 106, 126, 129, 130, 131, 133, 134, 137, 138, 139, 140, 169, 175, 179, 182, 184, 185, 188, 191, 196, 198, 200, 202, 203, 204, 205, 206, 211, 213, 216, 217
germination, 144, 147, 148, 150, 151, 164, 165, 166, 168, 169, 170, 182
glutathione, 101, 155, 163, 185
growth, 44, 101, 140, 144, 146, 153, 164, 165, 167, 168, 185, 191, 192

H

harmful effects, 110, 125, 187, 188

health, 1, 5, 6, 10, 13, 14, 15, 18, 22, 24, 27, 31, 32, 35, 43, 52, 53, 55, 57, 63, 96, 110, 134, 135, 162, 170, 174, 176, 184, 187, 190, 199, 201, 215, 219
heavy metals, 144, 151, 153, 162, 191, 202
Hi-C, 76, 77
histone, 75, 83, 87, 88, 216
histones, 36, 39, 75, 88
Hordeum vulgare L., 168, 187, 188, 189, 191, 192
HpaII, 86, 92
human colon cancer cells, 204, 206
human health, 52, 174, 187
hydrogen, 4, 134, 155
hydrogen peroxide, 4, 134, 155
hydrolysis, 174, 189
hydroquinone, 203, 204, 205, 206, 207, 208, 210, 211, 212, 213, 214, 215, 216, 217, 218, 219
hyperglycemia, 7, 10
hypermethylation, 89, 94, 96, 97
hyperosmolality, 204, 213, 219
hyperthermia, 30, 199

I

identification, 2, 27, 138
image analysis, 39, 115, 135, 136, 177, 192, 210
in vitro, 3, 4, 19, 33, 37, 41, 43, 49, 54, 56, 58, 59, 60, 62, 92, 94, 105, 107, 109, 110, 111, 123, 124, 125, 126, 129, 130, 131, 132, 134, 135, 136, 137, 139, 140, 169
In Vitro Applications, 4
in vitro exposure, 94, 139
in vivo, 19, 21, 33, 34, 37, 54, 59, 79, 80, 91, 107, 110, 111, 126, 131, 140, 206
incubation time, 127, 130, 209
induction, 14, 32, 64, 107, 110, 111, 128, 138, 140, 146, 152, 154, 185
industrialization, 18, 143

industry, 6, 14, 22, 24, 26, 31, 40
inhibition, 27, 32, 145, 146, 151, 152, 153, 170, 185, 193, 213, 216
integrity, 30, 47, 107, 146, 206
isolation, 12, 18, 25, 106, 115
issues, 35, 37, 52, 80, 133, 183, 219

K

kinetics, 66, 67, 68, 69, 75, 79, 80, 81, 102, 205
kinetics of the comet formation, 67, 68, 69, 80

L

lead, 5, 6, 15, 22, 25, 27, 40, 42, 47, 59, 85, 88, 109, 130, 146, 200, 211
leaves, 143, 144, 147, 148, 149, 150, 151, 152, 153, 154, 156, 157, 158, 159, 160, 161, 163, 166, 170, 173, 175, 177, 182, 184, 185, 188, 189, 191, 192, 193, 199, 204
lesions, 3, 36, 86, 97
leukocytes, 3, 5, 7, 14
light, 20, 43, 55, 65, 154, 163, 191
liquid biopsy, 89, 97
liver, 4, 5, 10, 17, 42, 61, 63, 204
loop anchors, 74
loop size distribution, 75, 77, 79
lung cancer, 30, 46, 49, 97
lymphocytes, 3, 6, 7, 8, 9, 10, 11, 13, 15, 17, 21, 23, 24, 25, 26, 27, 29, 30, 31, 32, 34, 43, 57, 78, 106, 108, 110, 111, 112, 115, 116, 120, 121, 122, 123, 124, 125, 126, 127, 129, 131, 132, 133, 134, 135, 137, 139, 141
lysis, 19, 21, 65, 66, 74, 75, 79, 82, 92, 93, 107, 114, 118, 119, 176, 209

M

mammalian cells, 10, 11, 21, 30, 32, 36, 39, 52, 55, 60, 79, 107, 140, 186, 205
manufacturing, 11, 26, 34
McrBC, 86, 93
measurement, 7, 75, 107, 149, 193
membranes, 2, 36, 156, 174
metabolism, 39, 109, 168, 218
metabolites, 174, 189, 210, 212, 216
methylation, 85, 86, 87, 88, 90, 91, 92, 93, 94, 97, 98, 99
methylation-specific restriction endonuclease, 92
Mg^{2+}, 153, 209
micronucleus, 2, 9, 15, 28, 59, 110, 111, 126, 133, 136, 181
microorganisms, 172, 174, 206
microscope, 2, 19, 20, 36, 92, 109, 114, 119, 149, 177, 180, 192, 193, 195, 210
microscopy, 17, 36, 92, 107, 168, 176
migration, 7, 20, 21, 26, 31, 65, 66, 67, 69, 70, 71, 72, 73, 78, 108, 109, 176, 182, 198
models, 4, 45, 47, 57, 59, 138, 183, 199
modifications, 36, 87
molecules, 37, 164
MspI, 86, 92
mutagenesis, 11, 12, 175, 196
mutations, 89, 107, 193

N

NaCl, 75, 114, 115, 119, 207, 209, 212, 213, 214, 219, 220
nanomaterials, 35, 36, 37, 39, 40, 44, 45, 48, 51, 53, 54, 55, 56, 58, 59, 60, 62, 63
nanoparticles, 35, 50, 52, 53, 54, 55, 56, 57, 58, 59, 60, 61, 62, 63, 64, 105, 126, 133, 138

neonicotinoid insecticide, 106, 110, 120, 132, 135, 136, 138, 141
nervous system, 110
neurotoxicity, 44, 61, 134
neutral, 3, 7, 19, 24, 65, 66, 67, 79, 105, 107, 108, 111, 119, 120, 122, 123, 124, 125, 126, 127, 130, 145, 167, 190
nuclei, 66, 71, 74, 79, 108, 149, 168, 182, 193, 196, 198, 201, 214
nucleic acid, 193
nucleus, 81, 82, 109, 144, 178, 182, 188, 196, 198
nutrient, 144, 191, 192, 207
nutritional studies, 6

O

occupational genotoxicity, 1
occupational groups, 19, 22
occupational health, 15, 31, 32
occupational studies, 5, 22
occupational toxicology, 17, 18, 19, 22, 24, 27
OECD, 22, 31, 37, 60, 126
organ, 109, 144, 145, 163, 176
organic solvents, 32, 174, 189
organism, 109, 169, 175, 189, 212
organs, 12, 16, 40, 42, 125, 145, 147, 191
oxidation, 14, 33, 36, 50, 101, 155, 205, 207, 215, 216
oxidative damage, 5, 8, 10, 11, 41, 49, 56, 108, 164
oxidative stress, 7, 10, 13, 15, 41, 53, 55, 58, 59, 61, 62, 64, 101, 129, 134, 137, 144, 153, 155, 156, 162, 163, 164, 166, 167, 168, 182, 184, 185, 200, 205
oxide nanoparticles, 55, 61
oxygen, 146, 163, 164

P

pathogenesis, 12, 87

pathways, 49, 86, 215, 217
Penicillium chrysogenum var.
  halophenolicum, 203, 218, 219
peripheral blood, 7, 8, 9, 15, 52, 57, 120,
  121, 122, 123, 124, 125, 129, 132, 134,
  141
peripheral blood mononuclear cell, 8, 52
pesticide, 5, 11, 23, 105, 125, 127, 130, 134,
  135, 137, 140, 171, 172, 173, 174, 175,
  176, 181, 183, 188, 189, 190, 192
pests, 171, 172, 173, 188, 189
pH, 21, 66, 67, 79, 113, 114, 115, 118, 119,
  144, 145, 147, 148, 149, 176, 177, 178,
  192, 193, 207, 209
pharmaceuticals, 22, 46, 89, 96, 105, 135
phenol, 205, 207, 215, 218
phenolic, 22, 155, 203, 204, 206, 211, 218,
  219
phenolic compounds, 22, 155, 211, 218, 219
phosphate, 47, 112, 141, 146, 148
photosynthesis, 145, 153, 154, 167
physicochemical properties, 35, 37, 46, 50,
  52
plant growth, 143, 145, 147, 150, 162
plants, 22, 24, 30, 36, 109, 144, 145, 150,
  153, 154, 155, 160, 163, 164, 165, 166,
  167, 169, 170, 172, 173, 174, 175, 176,
  181, 184, 185, 187, 189, 190, 191, 192,
  193, 196, 197, 199, 200, 204, 219
platform, 94, 95, 100, 136
point mutation, 107, 184, 199
poison, 174, 205, 216
pollutants, 18, 22, 106, 132, 173, 174, 187,
  210
pollution, 48, 139, 145, 217
polycyclic aromatic hydrocarbon, 5, 25, 34,
  217
polymerization, 71, 146
potassium, 140, 165, 207
preparation, 22, 49, 113, 120, 121, 124, 176
prevention, 87, 88
principles, 29, 65, 79, 81, 83, 108, 131, 183

prognosis, 86, 90, 94
promoter, 88, 89, 96, 97
protection, 15, 42, 109, 173
protective mechanisms, 182, 198
protein-protein interactions, 74, 83
proteins, 49, 66, 74, 75, 76, 88, 90, 155,
  156, 176
proteome, 166

Q

quantification, 37, 100, 144, 187

R

radiation, 14, 18, 22, 23, 25, 31, 32, 34, 60,
  70, 79, 186
radicle, 150, 168
reactive oxygen, 43, 55, 98, 128, 146, 155,
  162, 164, 166, 205, 211, 216
reactivity, 41, 43, 44, 48, 52, 61
recovery, 42, 128, 134
regions of the world, 171, 172
regression, 146, 150, 153, 159
relevance, 35, 61, 164
remediation, 166, 204, 210, 212, 214, 219
researchers, 87, 127, 128, 129, 146, 151,
  152, 158
residues, 21, 97, 134, 199
resistance, 41, 70, 90, 144, 167, 169, 170,
  171, 175, 176, 188, 205
response, 14, 41, 49, 52, 57, 58, 62, 94, 145,
  154, 158, 160, 161, 162, 163, 191, 206
reversibility of DNA migration, 70
reversibility of the DNA exit, 71
risk, 2, 6, 14, 23, 25, 27, 40, 43, 90, 105,
  126, 132, 134, 139, 143, 175, 184, 187,
  196
risk assessment, 14, 105, 132
risks, 29, 32, 141, 189
room temperature, 93, 113, 114, 115, 198,
  209

Index

root, 144, 145, 147, 150, 162, 163, 164, 167, 168, 169, 173, 174, 176, 182, 184, 185, 191, 197, 198, 202
root growth, 144, 145, 146, 169, 174
root system, 144, 147, 182, 191
roots, 144, 146, 150, 160, 163, 169, 170, 174, 184, 191, 202
routes, 40, 46

S

scatter plot, 212, 213
seed, 147, 150, 151, 168, 169, 170
seedlings, 144, 148, 162, 168, 170, 171, 172, 178, 179, 185, 188, 194, 196, 198
sensitivity, 5, 19, 27, 91, 92, 94, 95
sepsis, 5, 10, 11, 15
serum, 8, 89, 90, 96, 100, 112, 115, 116, 208
showing, 50, 178, 207, 214
silica, 24, 25, 28, 46, 47, 56, 59, 64
silver, 20, 40, 52, 54, 55, 56, 58, 59, 60, 61, 63
single cell gel electrophoresis (SCGE), 17, 19, 175, 196
SiO2, 45, 56, 59
sister chromatid exchange, 2, 29, 107, 133, 167, 191
sodium, 24, 74, 115, 119, 147, 148, 209
software, 109, 115, 119, 149, 193, 208, 209, 210
solution, 21, 68, 69, 75, 92, 93, 112, 113, 114, 115, 118, 119, 144, 147, 166, 171, 177, 188, 191, 192, 209
species, 4, 14, 43, 55, 98, 128, 146, 152, 153, 154, 155, 158, 160, 163, 164, 165, 166, 168, 169, 171, 202, 205, 210, 211
sperm, 3, 8, 11, 12, 17, 21, 24, 26, 27, 29, 30, 33
stability, 45, 46, 51, 85, 88
standard deviation, 152, 154, 156, 157, 159
stomata, 151, 152, 164, 168

stress factors, 144, 156, 173, 175, 190
structure, 36, 37, 41, 48, 49, 50, 67, 76, 83, 86, 88, 108, 145, 146, 164, 171, 175, 178, 196, 212
sulfate, 174, 217
sunflower, 143, 144, 147, 148, 149, 150, 151, 152, 153, 154, 156, 157, 158, 159, 160, 161, 193
supercoiled loops, 3, 67, 70, 72, 92
supplementation, 6, 8, 13, 15
susceptibility, 25, 31, 101
synthesis, 52, 146, 153

T

tail length, 20, 27, 67, 75, 109, 119, 128, 158, 160, 182, 196, 204, 210, 211, 213
target, 21, 35, 42, 88, 105, 110, 125, 144, 145, 150, 172, 173, 174, 175, 187, 188, 189, 190
testing, 33, 55, 56, 60, 80, 86, 90, 105, 109, 110, 117, 126, 130, 135, 140, 190
therapy, 8, 14, 48, 49, 51, 87, 90
thiacloprid, 106, 109, 110, 111, 112, 113, 116, 120, 121, 122, 123, 124, 127, 128, 130, 132, 134, 135, 139
tissue, 3, 21, 40, 48, 49, 90, 95, 147, 162, 163, 185
tobacco, 18, 27, 28, 30, 151, 163, 169, 183, 184, 190, 191, 199, 200
toxic effect, 37, 169, 175, 190, 204
toxicity, 37, 41, 43, 49, 50, 54, 56, 61, 135, 136, 141, 144, 145, 147, 150, 151, 152, 153, 155, 158, 160, 162, 164, 165, 166, 167, 168, 169, 170, 177, 182, 189, 191, 200, 202, 203, 212, 214, 215, 217, 218, 220
toxicology, 11, 16, 18, 19, 24, 27, 33, 59, 140, 141, 175
toxicology studies, 18, 19, 27
transcription, 78, 81, 83, 85, 88, 205
transformation, 49, 62, 89, 174, 188, 205

triggers, 53, 205
Triticum aestivum L., 171, 172, 173, 176, 185
tumor, 7, 31, 86, 88, 90, 96, 97
tumor cells, 7, 86, 89
tumor development, 89
tumors, 23, 89
Turkey, 1, 6, 17, 23, 24, 26, 32, 85, 100, 143, 147, 171, 187, 199
turnover, 89
type 2 diabetes, 14

V

validation, 12, 21, 34, 95, 107, 126, 135, 140, 201
variables, 27, 210
varieties, 158, 165
vertebrates, 190
viability, 106, 117, 120, 121, 122, 123, 124, 125, 129, 205, 208, 210, 211, 212
vulnerability, 14

W

waste, 22, 30
waste disposal, 22, 30
water, 4, 18, 109, 112, 114, 117, 144, 146, 151, 165, 172, 173, 174, 176, 184, 188, 189, 200, 208, 210
water resources, 173, 188
workers, 5, 6, 9, 10, 11, 13, 14, 15, 17, 18, 22, 23, 24, 25, 26, 28, 29, 30, 31, 32, 33, 34, 42
workplace, 18, 22, 27
worldwide, 109, 143

Y

yield, 87, 143, 146, 165, 173, 187, 188

Z

zinc, 40, 43, 55, 57, 59, 61, 166, 167
ZnO, 56, 58, 60, 64

Related Nova Publications

BIOCHEMISTRY LABORATORY MANUAL FOR UNDERGRADUATE STUDENTS

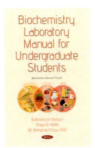

AUTHORS: Buthainah Al Bulushi, Raya Al-maliki, and Musthafa Mohamed Essa, Ph.D.

SERIES: Biochemistry Research Trends

BOOK DESCRIPTION: This laboratory manual has been designed for nutrition students for a better understanding of the lab assessments including biochemistry and food chemistry lab assessments. This manual includes both qualitative and quantitative analyses of some of the macro and micronutrients.

ONLINE BOOK ISBN: 978-1-53614-967-8
RETAIL PRICE: $0

AMYLASES: PROPERTIES, FUNCTIONS AND USES

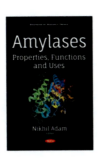

EDITOR: Nikhil Adam

SERIES: Biochemistry Research Trends

BOOK DESCRIPTION: *Amylases: Properties, Functions and Uses* opens with an analysis of the methods commonly used for the immobilization of amylase on particles, the effect that the processes of adsorption and covalent immobilization have on the activity and stability of the enzyme, as well as on its stability and reusability.

SOFTCOVER ISBN: 978-1-53614-993-7
RETAIL PRICE: $82

To see a complete list of Nova publications, please visit our website at www.novapublishers.com

Related Nova Publications

HORSERADISH PEROXIDASE: STRUCTURE, FUNCTIONS AND APPLICATIONS

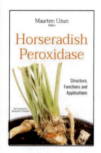

EDITOR: Maarten Uzun

SERIES: Biochemistry Research Trends

BOOK DESCRIPTION: In this compilation, the authors discuss the commercial source for the enzyme horseradish peroxidase, the tuberous roots of the horseradish plant which is native to the temperate regions of the world. Horseradish peroxidase is an oxidoreductase belonging to the highly ubiquitous group of peroxidases, indicating that this enzyme came into existence in the early stages of evolution and has been conserved thereafter.

SOFTCOVER ISBN: 978-1-53615-912-7
RETAIL PRICE: $95

HEMAGGLUTININS: STRUCTURES, FUNCTIONS AND MECHANISMS

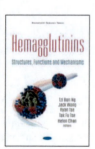

EDITORS: Tzi Bun Ng, Jack Wong, Ryan Tse, Tak Fu Tse, and

SERIES: Biochemistry Research Trends

BOOK DESCRIPTION: Hemagglutinins refers to glycoproteins which bring about agglutination of erythrocytes or hemagglutination. Hemagglutination can be used to identify surface antigens on erythrocytes (with known antibodies) and, hence, the blood type of an individual.

HARDCOVER ISBN: 978-1-53615-708-6
RETAIL PRICE: $230

To see a complete list of Nova publications, please visit our website at www.novapublishers.com